PROCEEDINGS
of the
FIRST DONEGANI SCIENTIFIC WORKSHOP
on
STRATEGIES FOR COMPUTER CHEMISTRY

PROCEEDINGS
of the
FIRST
DONEGANI SCIENTIFIC WORKSHOP
on
STRATEGIES FOR
COMPUTER CHEMISTRY

October 12-13, 1987

Edited by

CAMILLO TOSI

Istituto Guido Donegani, Novara, Italy

KLUWER ACADEMIC PUBLISHERS

DORDRECHT / BOSTON / LONDON

Library of Congress Cataloging in Publication Data

```
Donegani Scientific Workshop (1st : 1987)
   Proceedings of the "First Donegani Scientific Workshop" on
 strategies for computer chemistry : October 12-13, 1987, Istituto
 Guido Donegani, Novara, Italy / edited by Camillo Tosi.
      p.   cm.
   Includes bibliographies and index.
   ISBN 9027728321
   1. Chemistry--Data processing--Congresses.    I. Tosi, Camillo.
 II. Istituto di ricerche Donegani di Novara.   III. Title.
 QD39.3.E46D66 1987
 542'.8--dc19                                              88-25224
                                                               CIP
```

ISBN 90-277-2832-1

Published by Kluwer Academic Publishers,
P.O. Box 17, 3300 AA Dordrecht, The Netherlands.

Kluwer Academic Publishers incorporates
the publishing programmes of
D. Reidel, Martinus Nijhoff, Dr W. Junk and MTP Press.

Sold and distributed in the U.S.A. and Canada
by Kluwer Academic Publishers,
101 Philip Drive, Norwell, MA 02061, U.S.A.

In all other countries, sold and distributed
by Kluwer Academic Publishers Group,
P.O. Box 322, 3300 AH Dordrecht, The Netherlands.

All Rights Reserved
© 1989 by Kluwer Academic Publishers
No part of the material protected by this copyright notice may be reproduced or
utilized in any form or by any means, electronic or mechanical
including photocopying, recording or by any information storage and
retrieval system, without written permission from the copyright owner.

Printed in The Netherlands

CONTENTS

R. GIACCONI
    Preface 1

A. COLLINA
    Welcome Address 3

C. TOSI
    The Why's of this Workshop 5

KJ. RASMUSSEN
    Development of Potential Energy Functions for Use in
    Conformational Analysis 13

F. AVBELJ, J. MOULT, D.H. KITSON*, M.N.G. JAMES and A.T. HAGLER
    Theoretical Studies of the Energetics and Dynamics of
    the Aqueous and Ionic Environment about Proteins:
    Crystals of Streptomyces Griseus Protease A 31

F. BERNARDI*, A. BOTTONI, J.J.W. MC DOUALL, M. OLIVUCCI, M.A. ROBB, G. TONACHINI and A. VENTURINI
    Bond Lengths in Transition Structures and Intermedia-
    tes of Cycloaddition Reactions 33

J. M. ANDRÉ
    Some Aspects of Computational Polymer Quantum Chem-
    istry 45

P. FANTUCCI*, V. BONAČIĆ-KOUTECKÝ and J. KOUTECKÝ
    Ab Initio Configuration Interaction Study of Elec-
    tronic and Geometric Structure of Alkali Metal Clus-
    ters 79

D.P. DOLATA
    WIZARD: Artificial Intelligence and Conformational
    Analysis 93

G. ZERBI
    State of the Art in Vibrational Dynamics of Large
    Molecules 109

G.B. BACHELET
    Density Functional Theory and First-Principles Pseu-
    dopotentials: Two Important Tools in Solid-State
    Theory 119

M. MARSILI
    Autodeductive Modeling and Optimization in Chemo-
    metrics 161

R. SCORDAMAGLIA* and L. BARINO
   Statistical Distribution of Molecular Conformations and Its Application in QSAR Research 179

L. BARINO* and R. SCORDAMAGLIA
   Molecular Chain Flexibility and Phase Transitions in Polymers 193

R. MOLL
   General Aspects of Computer-Aided Synthesis Planning   213

R. FUSCO*, L. CACCIANOTTI and C. TOSI
   Detection and Structural Description of the Deepest Minima in a Potential Energy Hypersurface 227

G. RANGHINO
   Monte Carlo Simulation of the Solvation of a Ribonucleotide 245

Round Table Discussion on
   The Organization of a Molecular Modeling Group in a Chemical Industry 251

AUTHOR INDEX 265

PREFACE

The initial idea for this series of "Donegani Scientific Workshops" came in discussions with the Scientific Council of the Istituto Donegani directed to explore ways and means whereby research could be strengthened and the usefulness to the Montedison Corporation increased. It is important to realize the special niche occupied by Istituto Donegani in the Montedison family of companies. It is the central research facility of the corporation and as such it is charged both with the preservation, nurturing and augmentation of a basic cultural heritage and with its application to specific design or production problems arising through the companies of the group from market requirements.

We at Donegani believe that we can best respond to these dual roles by providing a small expert core of highly qualified scientists who are involved in basic research and whose activities are directed to the acquisition of the competence required by strategic long-term goals of the Montedison Corporation. This group interacts and merges (from time to time) with project-oriented research teams focused on the solution of problems of immediate tactical importance.

Under the stimulus of market and financial constraints, it is possible for short periods of the corporate life to ignore the need to maintain a strong intellectual basis for the day-to-day activities. However, it is obvious that in the long term, such short-sighted policies directed to maximize the perceived utility of the research institution will result in a level of competence and expertise insufficient to provide strong scientific and technical support for corporate strategic and even tactical decisions.

Given the scope and importance of Montedison activities in the pharmaceutical, chemical, agricultural and materials sectors on the Italian and European scene, Istituto Donegani represents an important corporate and national resource. The Institute has a particular opportunity and responsibility, building on its past and current performance, to aim for a level of excellence, in chosen discipline areas, equal to that of the best research laboratories in the world.

In pursuing this objective we have, in the last few years, engaged in effort to strengthen our core expertise as well as our applied research performance through a number of coordinated initiatives. This scientific workshop is intended to furnish a forum to scientists within and without the Institute, for the development of ideas, in fields in which Donegani is interested; to foster a comprehensive, highly competent and up-to-date view of the state of the art in each field; and to encourage, when appropriate, collaborative, joint or complementary research efforts with other research institutions in the industrial, governmental or academic field.

Publication of these proceedings is aimed to serve experienced practitioners in the field as well as young researchers. In perusing the draft of this volume, I have become convinced that we are largely achieving these aims thanks to the enthusiastic participation by Dr. Camillo Tosi and his group and the many colleagues from all over the world who have participated in the workshop with research contributions and discussions. To all of them, my sincere gratitude.

Riccardo Giacconi
Chairman of Istituto Donegani

WELCOME ADDRESS

I want to express my appreciation to all those who have come here, and in particular to all those who will actively collaborate towards the success of the workshop, either by giving lectures or by taking part in the discussions.

The presence of such a large and qualified audience undoubtedly proves that the subject matter of this workshop is of great interest and actuality. The reasons for this are many and depend both on the importance acquired by computer-aided chemistry and the increasing impact it has in chemical research.

I am particularly happy to organize here at Istituto Donegani this meeting on "Strategies for Computer Chemistry", since it reminds me of the period I spent, about twenty years ago, at the Montedison Computational Centre in Milan. This experience has enabled me to perceive not only the giant steps taken by computer chemistry over the last two decades, but also its amazing potential for the coming years. The role I cover puts me in an ideal position to transform this perception into operative decisions. The personnel of the Molecular Modeling Department and of the Computational Centre of the Institute can witness the sensitivity I have always had, and shall continue to have, for their needs.

"The computer has changed many of the ways of science and engineering. It has changed the way we think and the way we approach problem solving.... The computer is converting traditional chemical discovery into a mathematical science, stretching the capacity of many old-school chemists to adapt" [1]. I am quite sure that the possibilities of developemnt for computer-aided chemistry are limited only by the imagination and creativity of those involved in this field.

I do not want to take any more time from your works. Let me conclude by wishing you all the best for your participation in this workshop and, more generally, for your scientific activity.

Amilcare Collina
Managing Director, Istituto Donegani

[1] From "Computer-Aided Chemistry: New Routes to Tomorrow's Drugs and Chemicals" (R.L. Davidson, Project Editor), Technical Insights Inc., Fort Lee, New Jersey (1986), p.10.

THE WHY'S OF THIS WORKSHOP

Camillo Tosi

Istituto G. Donegani
Via G. Fauser 4, I-28100 Novara

The number of symposia, congresses, conferences and workshops that take place every year all over the world is nowadays so large, that anyone who decides to organize a meeting must give convincing justifications to the colleagues who have agreed to participate in it that they will not be wasting their time. It is particularly important that they do not have the troublesome feeling of "increasing the tower of Babel by adding another opportunity for fragmentation of the scientific community" [1].

Let me share with you the reasons which have convinced me that this workshop is not a duplicate of others you may have already attended in the recent past. To begin with, when and how did the idea of a workshop like this first develop? As the Gospel teaches us, one should render unto Caesar the things that are Caesar's, and unto God the things that are God's. Accordingly, it is fair to tell you that the first suggestion of this workshop was made by Prof. R. Giacconi, the well known astrophysicist who, about one year ago, was appointed as Chairman of the Guido Donegani Institute. The driving ideas beneath his proposal are essentially based on the following considerations.

Computer chemistry has become one of the most important activities in a chemical company, especially when this company, as is the case for Montedison, has multifaceted interests spanning over a wealth of different, though related, disciplines. However, computer chemistry is a young science, which is still blooming. As a consequence, there is the perennial risk that, like all things which have grown much too rapidly, there is a lack of balance in the preparation of people carrying out this activity. Some periodical check-ups with a very precise scope are thus needed: to highlight their strengths and limitations, and hopefully to take the correct decisions to consolidate the former and eliminate the latter. An efficient check-up is represented by the organization of a meeting with a number of outstanding experts in the various subfields, so that they can make a first-hand evaluation of their level of preparation.

This being the leading idea, I have tried to do my best to diagnose the state of health of our molecular modeling team. This workshop should give an independent validation of my diagnosis and help me to propose, if necessary, the most useful therapies.

As many of you may know, our immediate precursor is the Department of Chemical Computation established by E. Clementi thirteen years ago. This origin has led, so to say, to a certain bias in our specialization, which, at least initially, was predominantly formed by a single ab initio quantum chemical program, IBMOL. However, Clementi had a very brilliant intuition, the great value of which has now been realized by the scientific community, as I had the opportunity to verify at the 1987 World Congress of the Association of Theoretical Organic Chemists held in Budapest two months ago: namely, that quantum mechanical computations, though feasible in principle, were none the less much too time-consuming to be applicable to any point of configurational space, so that ab initio results had to be generalized in order to be useful for a complete description of the energy surface of molecules. The adopted generalization consisted in fitting some simple analytical expressions to the finite set of computed energies, these expressions being more reliable, the more accurate the selection of points for which ab initio computations had been performed. This strategy was applied both to intermolecular interactions (water with both amino acids [2,3] and the DNA bases [4]), and to intramolecular interactions (the sugar-phosphate-sugar fragment, regarded as the smallest unit representative of the conformational behaviour of polynucleotide chains [5-7]).

The use of analytical potentials introduced us into the world of molecular mechanics. One of the leading contributors to this area of research, our first speaker Kj. Rasmussen, perceived the importance of this approach in the parameterization of potential energy functions, and started a collaboration with us which dates back to late 1977 and has given rise for me to a not only cultural, but also human enrichment.

Another development of our field of interests followed on from another consideration, namely that, when studying a molecular energy surface, it is often more advantageous to obtain less accurate information on a large portion of the surface than more precise information on a single point or a very limited number of points. This led us to turn to semiempirical quantum mechanical methods, and the greatest attention was devoted to PCILO, which is three to four orders of magnitude faster than IBMOL, and is particularly reliable for angular conformational analysis of organic molecules. A useful version of this program (PCILOPT), allowing for the simultaneous optimization of up to 9 internal degrees of freedom - whether bond lengths, bond angles or internal rotation angles - was written here in the late Seventies by another of our speakers, P. Fantucci [8], and was widely applied to a variety of problems of practical concern [9-23].

Although IBMOL, PCILOPT and the CFF program given us by Rasmussen have been our strong points over the years, we have always followed the policy of developing new computer programs that not only responded to specific needs originating from our daily activity, but also gave us the possibility of exchanging them with programs written by colleagues of other computational centers in Italy and abroad. I do not wish to establish a list of merits among these self-generated programs [24-29], but let me give special mention to the FMFIT program [24], a pioneering work allowing a semi-automatic fitting of rotatable bonds in flexible molecules on rigid reference structures by user-defined "branches" within the two molecules, and to the GLOMIN program [27,30], based on an algorithm developed by Aluffi-Pentini, Parisi and Zirilli [31], which seems to be the best available today for detecting the global minimum of conformational energy surfaces both for the solidity of its mathematical foundation and the goodness of its numerical performance.

The library of programs existing in our Computer Center for studying the structure of molecules and its relationship to their properties has been called "Donegani Molecular Modeling System" (DMMS for short). It is interesting to follow the historical development of this package and its applications through its various presentations: the first at the Eger (Hungary) Symposium on Steric Effects in Biomolecules in 1981 [32], the second at the 1984 Montedison Chemistry Meeting [33], the third in a paper appearing in the January 1987 issue of "La Chimica e l'Industria" [34], on the occasion of the 50th anniversary of the Donegani Institute.

Up to now I have made an exposition, not exhaustive but sufficiently detailed, of our past and present work. Should our future work move along the same lines, I could conclude that we are on the safe side. But if we wish to extend our domain of knowledge, much has to be done, and the speakers at this workshop can give an effective contribution to our development. This leads me to give the reasons for the choice of topics of the workshop.

i) The investigation of mechanisms of chemical reactions with the "diabatic surface" formalism, much implemented by Prof. F. Bernardi [35], will certainly be of invaluable help to synthetic organic chemists, who will be able to learn theoretically the trend of a reaction by following the steepest-descent paths (in mass-weighted coordinates) connecting minima of potential energy surfaces (corresponding to reactants or products) and saddle points (corresponding to transition structures).

ii) The knowledge of the electronic structure of metal atom clusters has far-reaching consequences for an understanding of some basic properties of matter as well as for practical applications, in particular heterogeneous catalysis. Who could talk about this topic better than Prof. P. Fantucci, who has recently written a very good review on its theoretical aspects [36], and spent much time in one of the leading centers in the world, the Institute of Physical and Theoretical Chemistry of the Free University of Berlin?

iii) The theory of clusters has been simultaneously developed using the concepts and methods of theoretical chemistry and those of solid-state physics. This interplay of the two branches of science seems to be an anticipation of a phenomenon which could lead, in the near future, to a revolution in the approach to the investigation of molecular structures. As you know, solving the Schrodinger equation for a system of many interacting electrons subject to the electrostatic field of nuclei is a very formidable problem; essentially exact solutions are only possible for light atoms and small molecules through the configuration interaction approach. While chemists have used the Hartree-Fock formalism, solid-state physicists usually apply an alternative method, the density functional theory. This approach has obtained remarkable success in the prediction of structural properties of materials, a field of increasing interest in our Institute. A key ingredient in this respect is the recent introduction of ab initio pseudopotentials which describe in a rigorous way, as opposed to the usual empirical pseudopotentials, the interaction of valence electrons with the atomic core. Methods and representative applications will be reviewed by an expert in this field, Dr. G. Bachelet.

iv) Polymers play a major role in modern technology. There are close links between advances in theoretical calculations on polymers and advances in computational science, and the computer has been an essential tool in developing both an electronic theory of polymers and, more recently, a molecular engineering approach in the organic solid state: think, for example, of the problems of metallic conductivities and non-linear optical effects in organic polymers. One of the world leading authorities in the field of polymer quantum chemistry is Prof. J.-M. Andre, who will give a lecture on the aspects of computational polymer quantum chemistry.

v) One of the techniques which in the future will presumably have an increasing impact on computational chemistry is the application of expert system techniques. At the recent Budapest congress I just mentioned, Clementi predicted that commercial machines based on artificial intelligence will be put on the market in the next decade; and in my opinion he is in a position to make a reliable forecast. Our program for the immediate future includes collaboration with two groups from the university and the CNR in Milan on the use of expert system methods in the field of conformational analysis. A computer program of this kind, called WIZARD, already exists, although it is limited, as far as I know, to working within the domain of saturated acyclic hydrocarbons [37]. Its author, Dr. D.P. Dolata, is certainly the best suited person to introduce this subject, and to show how the use of symbolic logic and high-level reasoning can substitute strictly numerical methods in the exploration of the conformational space of molecules.

vi) While molecular mechanical methods give us a "static" picture of a molecule, it is often more important to study its dynamical behaviour. This is particularly true in the case of proteins: if these molecules, fundamental for life, were motionless, they could not work; it is their internal movements that are at the base of their activity. These motions can be analyzed by computer simulation through molecular dynamics techniques, to which essential contributions have heen made by A. Hagler and his collaborators of the Agouron Institute in La Jolla, California. One of them, Dr. D. Kitson, will speak to us about the molecular dynamics simulation of a protein crystal.

vii) Among the applications of molecular dynamics, a very important role is played by the study of molecular motions activated when energy is supplied to linear chains, such as for example organic polymeric materials, biological membranes, phospholipids and so on. Computational predictions on these systems can be compared with results obtained with optical, magnetic, mechanical and dielectric experiments, depending on the time scale of these motions. An outstanding contribution in this field is made by Prof. G. Zerbi, and I do believe that his talk on the vibrational dynamics of systems relevant in modern technology will have a most stimulating effect for many research activities currently carried out in our Institute.

viii) Finally, Dr. M. Marsili will give a talk on the industrial applications of computer chemistry. I thought it was a must for me to invite him here, since he is the leader of an Italian Project on Computer Chemistry, whose aim is the development, over the next few years, of a large system of artificial intelligence in chemistry, called SUPERNOVA. This system is user-friendly, largely autodeductive and multitasking, deals with four main topics (drug design, synthesis design, molecular modelling and process optimization) and is specifically oriented towards the solution of problems of particular interest for the chemical and pharmaceutical industry [38].

I hope that the aims of this workshop are now clear, and can thus conclude by expressing the wish that all of you may consider your participation useful and profitable. Thank you and good work.

References

[1] G.R. Marshall, J.G. Vinter and H.-D. Holtje, J. Computer-Aided Molec. Design 1, 1 (1987).

[2] E. Clementi, F. Cavallone and R. Scordamaglia, J. Am. Chem. Soc. 99, 5531 (1977).

[3] G. Bolis and E. Clementi, ibid. 99, 5550 (1977).

[4] R. Scordamaglia, F. Cavallone and E. Clementi, ibid. 99, 5545 (1977).
[5] O. Matsuoka, C. Tosi and E. Clementi, Biopolymers 17, 33 (1978).

[6] C. Tosi, E. Clementi and O. Matsuoka, ibid. 17, 51 (1978).

[7] C. Tosi, E. Clementi and O. Matsuoka, ibid. 17, 67 (1978).

[8] P.C. Fantucci and C. Tosi, Abstracts XVI Natl. Congress Ital. Assoc. Phys. Chem., 33 (1981).

[9] G. Castellani, R. Scordamaglia and C. Tosi, Gazz. Chim. Ital. 110, 457 (1980).

[10] F. Salvetti, A. Buttinoni, R. Ceserani and C. Tosi, Eur. J.

Med. Chem. **16**, 81 (1981).

[11] C. Tosi, L. Barino, G. Castellani and R. Scordamaglia, J. Mol. Struct. Theochem **87**, 315 (1982).

[12] C. Tosi and W. Saenger, Chem. Phys. Letters **90**, 277 (1982).

[13] C. Tosi and R. Hilgenfeld, Theor. Chim. Acta **62**, 29 (1982).

[14] C. Tosi, Abstracts XVII Natl. Congress Ital. Assoc. Phys. Chem., 35 (1982).

[15] C. Tosi, Nuovo Cim. D **2**, 15 (1983).

[16] Kj. Rasmussen and C. Tosi, Acta Chem. Scand. A **37**, 79 (1983).

[17] C. Tosi, L. Barino, P. Melloni and F. Salvetti, J. Mol. Struct. Theochem **104**, 135 (1983).

[18] C. Tosi, L. Barino and R. Scordamaglia, ibid. **106**, 241 (1984).

[19] C. Tosi, J. Comp. Chem. **5**, 248 (1984).

[20] C. Tosi, J. Mol. Struct. Theochem **110**, 23 (1984).

[21] C. Tosi, R. Fusco, G. Ranghino and V. Malatesta, ibid. **134**, 341 (1986).

[22] S. Penco, A. Vigevani, C. Tosi, R. Fusco, D. Borghi and F. Arcamone, Anti-cancer Drug Design **1**, 161 (1986).

[23] A. Bernardi, M.G. Beretta, V. Malatesta and C. Tosi, Chem. Phys. Letters **133**, 496 (1987).

[24] L. Barino, Computers & Chem. **5**, 85 (1981).

[25] R. Pavani and G. Ranghinn, ibid. **6**, 133 (1982).

[26] G. Castellani and R. Scordamaglia, ibid. **8**, 127 (1984).

[27] C. Tosi, R. Pavani, R. Fusco, F. Aluffi-Pentini, V. Parisi and F. Zirilli, Rend. Accad. Naz. Lincei, Classe Sci. Fis. Mat. Nat. **78**, 149 (1985).

[28] R. Fusco, L. Caccianotti and C. Tosi, Nuovo Cim. D **8**, 211 (1986).

[29] L. Barino and R. Scordamaglia, Chim. Ind. (Milan) **68** (11), 104 (1986).

[30] C. Tosi, "Theoretical Chemistry in Drug Design: A New Method of Global Optimization", Conference Papers Computer-Aided Molecular Design (Basel, 1985), IBC Techn. Serv. Ltd.

[31] F. Aluffi-Pentini, V. Parisi and F. Zirilli, J. Optim. Theory & Applic. **47**, 1 (1985).

[32] R. Scordamaglia, L. Barino, G. Castellani, G. Ranghino and C. Tosi, Abstracts Symposium on Steric Effects in Biomolecules (Eger, 1981), p. 22.

[33] C. Tosi, Atti II Giornata Chimica Montedison (Milano, 1984), p. 19.

[34] C. Tosi, R. Scordamaglia, L. Barino, G. Ranghino, R. Fusco and L. Caccianotti, Chim. Ind. (Milan) **69** (1-2), 68 (1987).

[35] F. Bernardi, J.J.W. Mc Douall and M.A. Robb, J. Comp. Chem. **8**, 296 (1987).

[36] J. Koutecky and P. Fantucci, Chem. Rev. **86**, 539 (1986).

[37] D.P. Dolata and R.E. Carter, J. Chem. Inf. Comput. Sci. **27**, 36 (1987).

[38] M. Marsili, E. Marengo, M. Salomone, C. Del Buono, F. Cammarata, G. Scavia and L. Caglioti, Chim. Ind. (Milan) **69** (5), 25 (1987).

Development of potential energy functions for use
in conformational analysis

Kjeld Rasmussen

Chemistry Department A
Building 207
The Technical University of Denmark
DK-2800 Lyngby

In this lecture I intend to deal with methods
and points of view which in my opinion are basic
to all molecular modelling. Properly parametrized
and optimised potential energy functions are
imperative if one wants a realistic model within
any of the approaches being advocated these years.
 Examples from work on alkanes, ethers, saccharides, coordination complexes and other classes of
substances are given.

BASIC CONCEPTS

The Consistent Force Field (CFF) is one way of modelling Nature. It originated in Israel[1] and was further developed and employed in several places, including Denmark[2,3].

Conformational analysis is of course basically an experimental field. As I see it, we endeavour to (1) identify conformers, (2) study their geometries, (3) measure their relative occurrences, (4) find paths of interconversion between them. We are not able to perform conclusive experiments in all cases, so we often have to resort to calculations based on well-tested models.

In the CFF approach we construct a parametrical representation of the potential energy within and between molecules.

From this representation, or potential energy function, we stive to calculate the value of any molecular quantity which is observable. The observable quantity may be a bond length, a torsional angle, a dipole moment, a frequency of vibration, a thermodynamic function at a specified temperature. We then compare the calculated with the observed value and improve the model till we get an acceptable agreement. Then we are in a position to calculate non-observed and non-observable quantities with some confidence, that is, to make true predictions, not only postdictions.

In our models, potential energy functions are composed of primary terms and secondary or correction terms. The primary terms represent bonded and non-bonded interactions, and the functions are, for example, parabolic or Morse for bonded; and Lennard-Jones or Buckingham, plus electrostatic, for non-bonded interactions. The secondary terms are essential, simply because a model based only on the primary is too primitive; we use the conventional harmonic angle and Pitzer functions. The various functions are shown in Figure 1.

$$\begin{cases} V_b = \sum \tfrac{1}{2} K_b (b-b_o)^2 \\ \text{bonds} \\ V_{n-b} = \sum \tfrac{1}{2} \left(\dfrac{A}{r^{12}} - \dfrac{B}{r^6} + \dfrac{e_i e_j}{r} \right) \\ \text{non-bonded interactions} \end{cases}$$

$$\begin{cases} V_\theta = \sum \tfrac{1}{2} K_\theta (\theta - \theta_o)^2 \\ \text{angles} \\ V_\phi = \sum \tfrac{1}{2} K_\phi (1 + \cos k\phi) \\ \text{single bonds} \end{cases}$$

Figure 1: Molecular potential energy functions

It is important for what follows to emphasize that b, r, θ and ϕ are variables, whereas all the other symbols are parameters. They are definitely not force constants, bond lengths or valence angles: K_b is not the force constant, b_o not the equilibrium bond length of any bond in any molecule; they are energy function parameters with dimensions of force constant and bond length. Force constants and equilibrium bond lengths are derived for individual bonds by calculation.

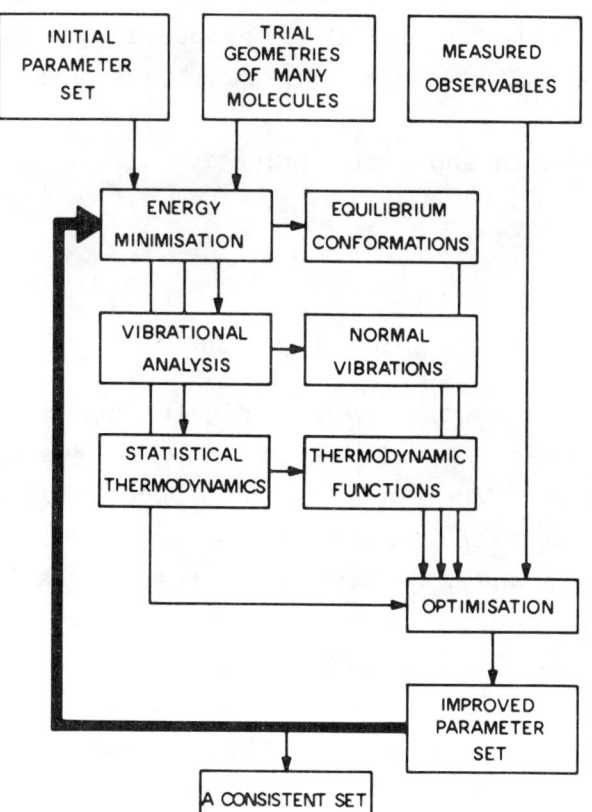

Figure 2: The CFF cycle.

SCOPE OF CALCULATIONS

The series of calculations performed under the CFF system in a typical application can be sketched as follows: (1) a sequence of related molecules is selected; (2) a potential energy function is composed; (3) parameters for the potential energy function are chosen from literature or are guessed; (4) initial conformations for all molecules are constructed; (5) these are varied until the energy is at a minimum; (6) any other property dependent of conformation is calculated; (7) calculated observables are compared with experimental data for selected properties; (8) parameter values of the potential energy functions are changed for better fit.

The CFF cycle is shown in Figure 2.

EXAMPLES: ALKANES AND SACCHARIDES

Step (8) above is the most intriguing and difficult, and much manpower has been invested in the development of a proper coding of the optimisation. Undoubtedly, the best solution has not been found, but the system works and gives good results, as shall be demonstrated a little later.

The optimisation is the ultimate scope: to produce a potential energy function which equally well reproduces a variety of data for a variety of molecules.

Of all the interactions mentioned, the non-bonded are both the most important when it comes to details of conformation and small differences in thermodynamic properties, and those that are most difficult to model satisfactorily. Many different functions have given almost equally good representations.

Figure 3 shows Buckingham functions[4] used quite successfully with saccharides[5,6], combined Lennard-Jones and

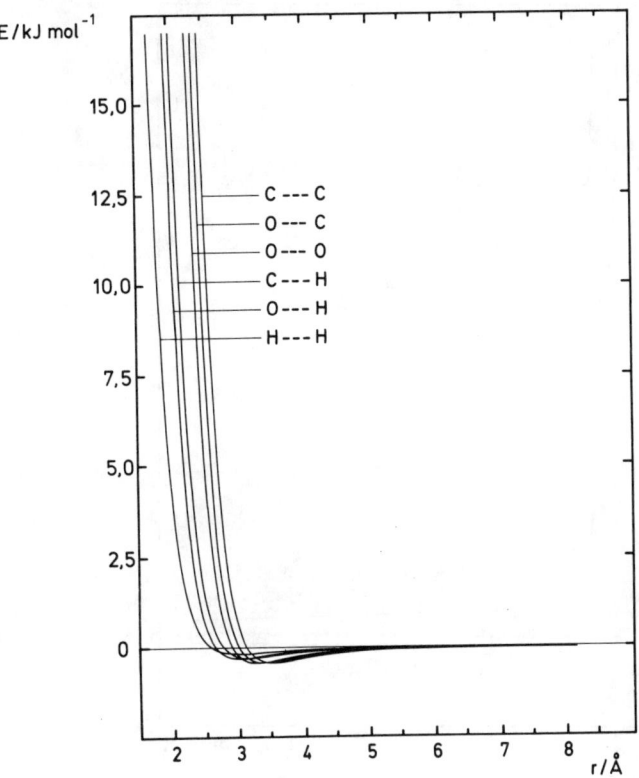

Figure 3: Buckingham functions selected for saccharides

electrostatic terms are shown in Figure 4 for use with alkanes[7] and in Figure 5 for use with ethers and alcohols[7] and with saccharides[8,9,10]. The same Lennard-Jones parameters are used in Figures 4 and 5.

These functions have proven satisfactory for postdiction of structures of glucose[4] and disaccharides[5,6] and of the anomeric equilibrium of glucose[4,10], and for similar predictions[8,9,10], for other saccharides.

It is noticeable that the functions were developed by trial-and-error[4,7], without proper optimisation, though admittedly with a sizeable use of working time.

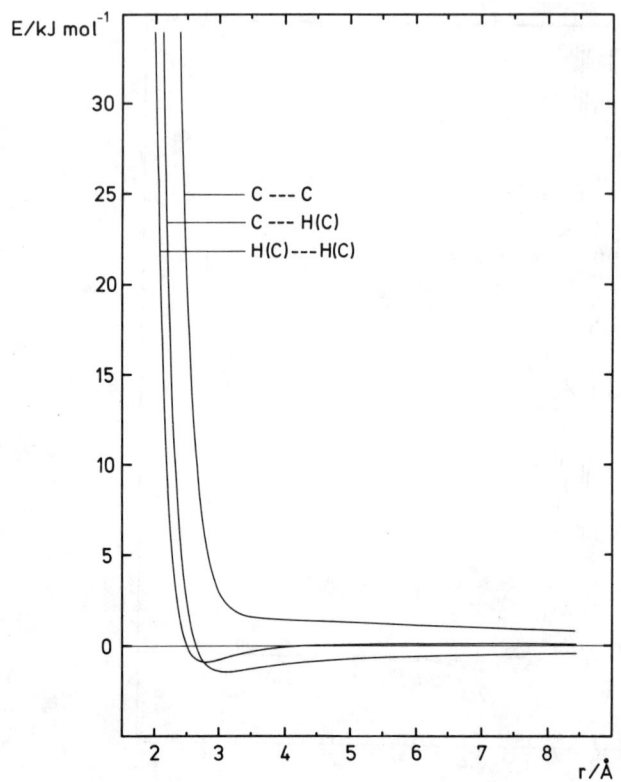

Figure 4: Lennard-Jones plus Coulomb functions for alkanes

The unusual forms of these functions: purely repulsive C---C, double-extremum C---H, deep-valleyed O---H interactions, are supported qualitatively by bond energy analysis (BEA)[11] of integrals resulting from ab initio calculations on saccharides[12], as exemplified in Figures 6 and 7.

EXAMPLES: COORDINATION COMPLEXES

Another group of examples of succesful application of non-optimized potential energy functions is found in our work on coordination complexes of Co(III) with diamines.

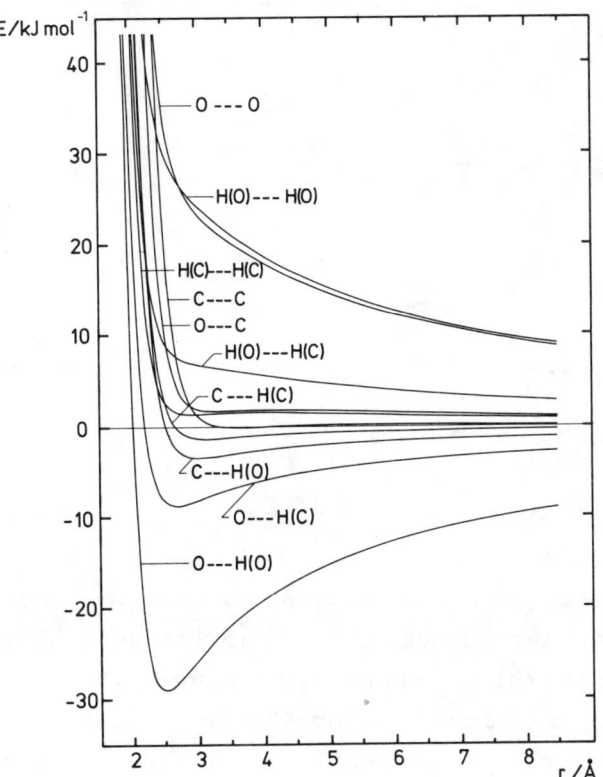

Figure 5: Lennard-Jones plus Coulomb for saccharides

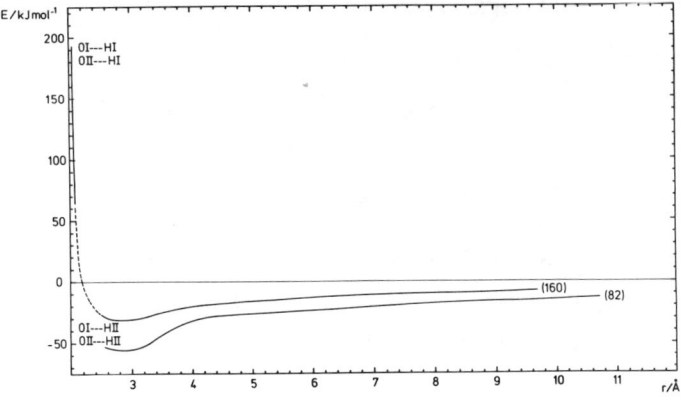

Figure 6: Non-bonded interaction between O and H from BEA

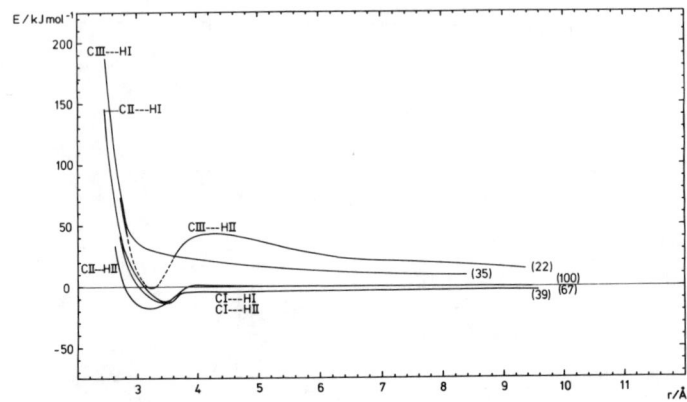

Figure 7: Non-bonded interaction between C and H from BEA

1,2-Ethanediamine (en for short) forms 5-membered, and 1,3-propanediamine (tn) forms 6-membered chelate rings. Remarkably good representations of their known structural[13,14] and thermodynamic[15] properties were obtained with non-optimised potential energy functions; accordingly, some confidence may be placed in predictions of structural[16,17] and thermodynamic[15,17] properties, and in the description[18] of the mechanism of ring-flipping in 5-membered chelate rings. Let it be noted, as well, that the low-frequency vibrations, so important for proper description of small differences in thermodynamic functions, are surprisingly well reproduced[19].

EXAMPLES: COMPOUNDS NOT USED IN FITTING

When developing potential energy functions whether by trial-and-error fitting or by optimisation, one should always test the results against data not used in the development. Also tests outside the range of obvious applicability should be carried out. One such example is found in our study of spiro compounds[20]. Here functions developed for unstrained alkanes and cycloalkanes are used on molecules with 3-, 4- and 5-membered rings. Much to our initial surprise, results were encouraging; particularly when comparing with the most recent experimental data. The structure of 4-rotane was published[21] after our manuscript was finished; the results are (calc, X-ray): C_{spiro}-C_{spiro}(1.517, 1.525), C_{spiro}-C_{other}(1.510, 1.502), C_{other}-C_{other}(1.518, 1.521). Considering the simplicity of the model and the amount of strain, the result is too good to be true; yet it is true.

OPTIMISATION

When it comes to optimisation, the algoritmic and numerical problems increase drastically, as mentioned before. Our chosen procedures are described in full elsewhere[3] and shall not be repeated here.

The question of which structure type to optimise on has not, in my opinion, been given sufficient attention in literature. I have summarized the various structure types before[3], and argued for the choice of r_z or r_α^o as the most appropriate for our purposes.

CRYSTALS

As stated above, non-bonded interactions are both highly important and difficult to model. Therefore we thought it necessary to turn to crystals, because their structure is almost exclusively dependent of non-bonded interactions.

It proved possible[22] to extend Williams' formalism[23,24] to our usual non-bonded functions, and a program is now operative which allows for convergent lattice summation and minimisation of all non-bonded energy terms, and for optimisation on unit cell dimensions and lattice energy. Some important properties of the program should be noted: it is possible to handle, in the same computation, and therefore within the same optimisation, flexible molecules in gaseous and crystalline phase, and molecular and ionic compounds. This program is probably the most versatile of those available. Unfortunately, it is not easy to run.

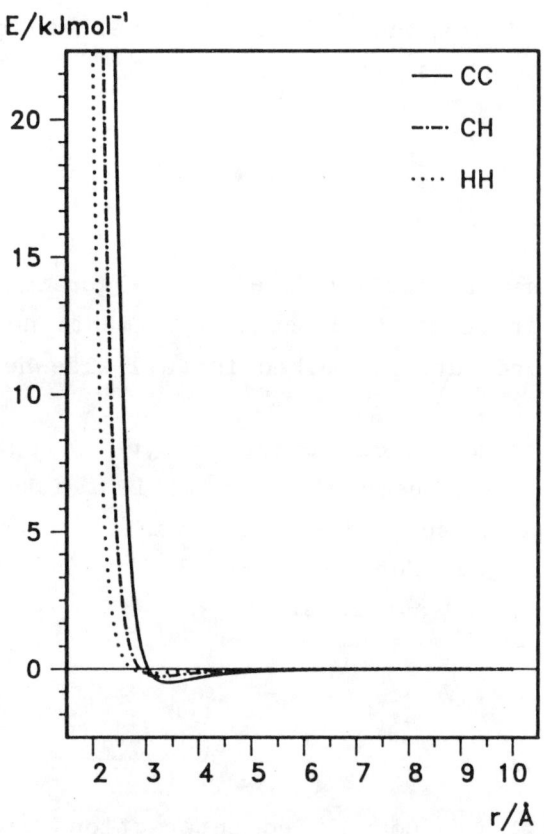

Figure 8: Optimised non-bonded functions, see text

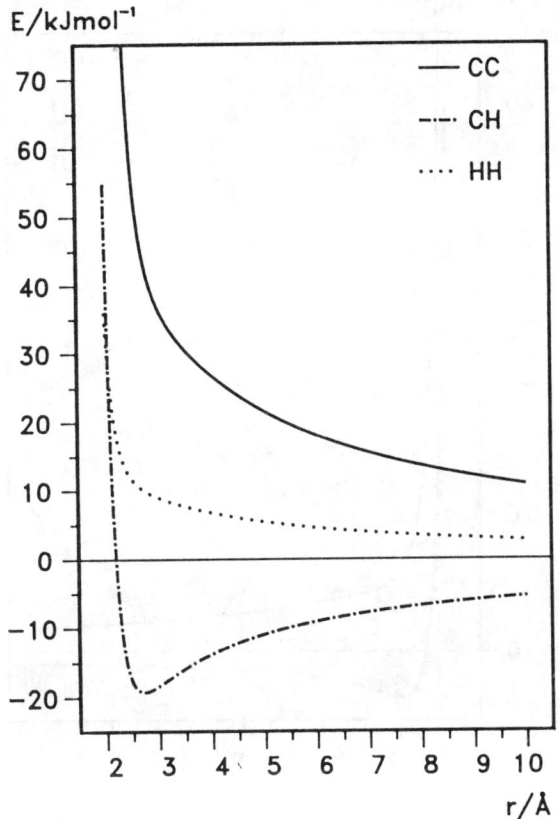

Figure 9: Optimised non-bonded functions, see text

EXAMPLES: NON-BONDED INTERACTIONS

Some principles of optimisation and properties of optimised potential energy functions will be illustrated by recording some of our experiences with a few very simple non-bonded functions. Their form is

$$V = \frac{A_i A_j}{r_{ij}^{12}} - \frac{B_i B_j}{r_{ij}^{6}} + \frac{e_i e_j}{r_{ij}}$$

and cannot be simpler.

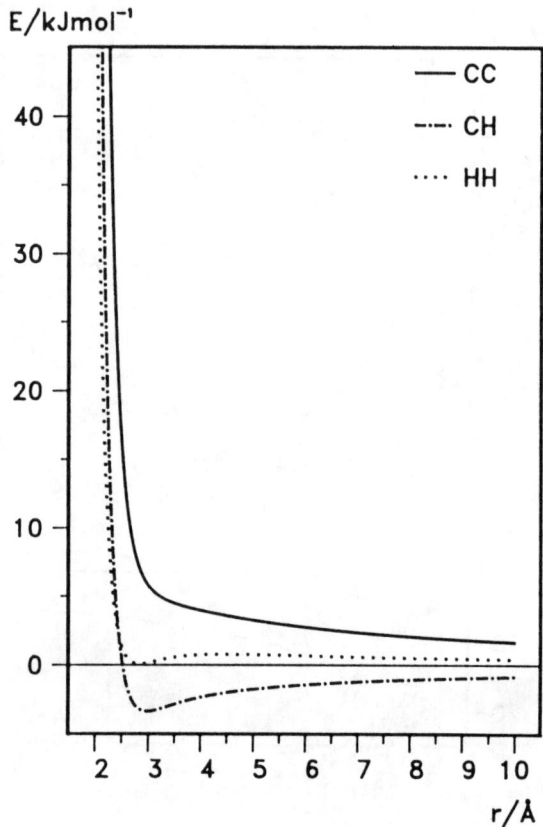

Figure 10: Optimised non-bonded functions, see text

After optimisation on small alkanes and cyclohexane, three different parameter sets gave very similar results. The differences between the functions are largely concentrated in the non-bonded interactions, as seen from Figure 8 (Lennard-Jones, no charges), 9 (same L-J, with charges) and 10 (different L-J, with charges).

When tested on compounds not included in the optimisations, all three give good, the two first excellent, representations of CCC angles ranging from 106 to 116°; entropy and heat capacity are well reproduced, and here the third function is the best.

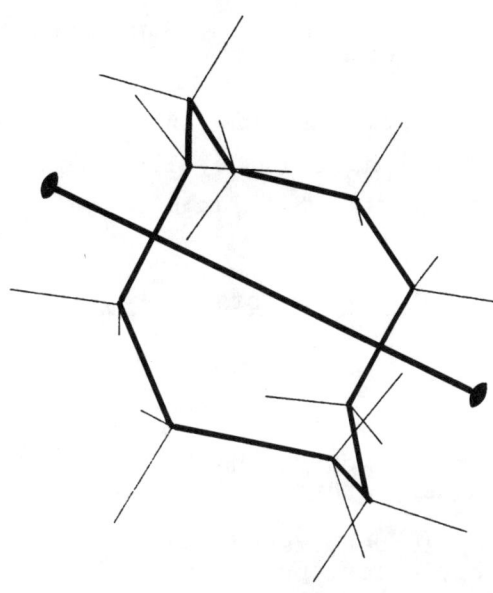

Figure 11: The BCB conformer of cyclodecane

Cyclopentane is a symmetric top, and its calculated frequency of pseudorotation is close to 0. Cyclodecane presented a problem to the non-optimised functions; now the structure of the BCB conformer, Figure 11, is reproduced to within experimental uncertainty; see Table 1.

Table 1: Cyclodecane BCB in optimised PEF's

	303	304	401	exp
$\theta_1\theta_4\theta_7\theta_9$	118	118	122	118
$\theta_2\theta_5\theta_6\theta_{10}$	115	115	117	114
$\theta_3\theta_8$	118	118	120	118
$-\phi_1\phi_4\phi_6-\phi_9$	54	54	54	55
$-\phi_2\phi_3\phi_7-\phi_8$	67	67	63	66
$-\phi_5\phi_{10}$	151	150	144	152

EXAMPLES: ETHERS

The most recent optimisations in our group have dealt with ethers, straight-chain and cyclic, some of which contain anomeric carbon atoms. See Table 2.

Table 2: Substance key

1 Ethane
2 Propane
3 Isobutane
4 Neopentane
5 Methylethylether, methyl end
6 Methylethylether, ethyl end
7 Dimethylether
8 Methyl-n-propylether, methyl end
9 Methyl-n-propylether, propyl mid
10 Methyl-n-propylether, propyl end
11 Diethylether
12 Bis(methoxy)methane, central
13 Bis(methoxy)methane, peripheral
14 Bis(methoxy)propane
15 Tetrakis(methoxy)methane, central
16 Tetrakis(methoxy)methane, peripheral
17 Cyclopentane
18 Cyclohexane
19 Tetrahydrofuran
20 Tetrahydropyran
21 1,4-Dioxan
22 1,3,5-Trioxan
23 2,4,6-Trimethyltrioxan
24 1,3-Dioxan
25 1,3-Dioxan, anomeric C

The work is not finished when this report is presented; but it is to be expected that little remains to be done. The preliminary results are shown in Fig. 12; plusses represent calculated and strokes experimental values.

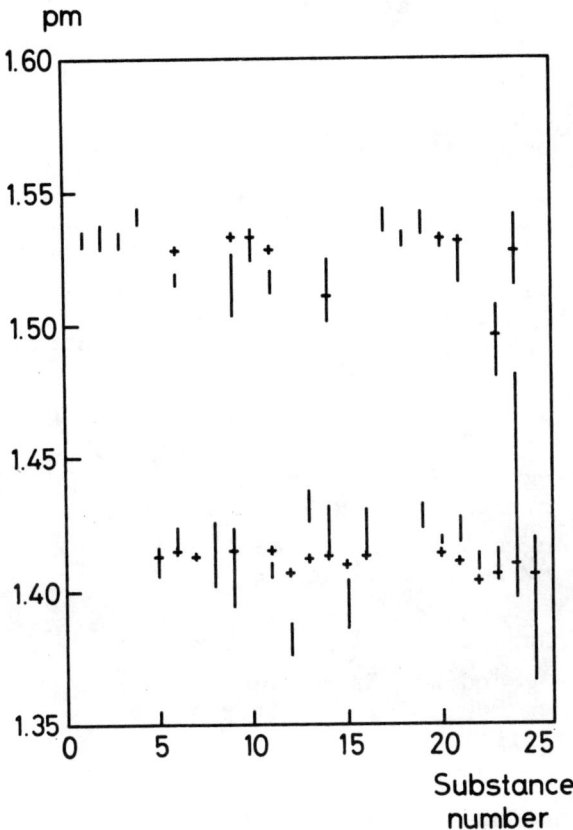

Figure 12: Optimisation on ethers, preliminary results

CONCLUSION

In conclusion, I want to state that (1) the CFF system is available to everybody[3]; it can be used for (2) conformational analysis of widely differing classes of compounds; (3) skill and know-how are required to handle it; (4) about 30 people contributed to the development of our version over almost 20 years; (5) our goal is to construct models with which we can calculate any property of any molecule. Se we have still a long way to go.

List of References

(1) S. Lifson and A. Warshel
J. Chem. Phys. 49 (1968) 5116-5129

(2) S.R. Niketić and Kj. Rasmussen
The Consistent Force Field: A Documentation. Lecture Notes in Chemistry, Vol. 3, Springer - Verlag (1977)

(3) Kj. Rasmussen
Potential Energy Functions in Conformational Analysis, Lecture Notes in Chemistry, Vol. 37, Springer - Verlag (1985)

(4) K. Kildeby, S. Melberg and Kj. Rasmussen
Acta Chem. Scand. A31 (1977) 1-13

(5) S. Melberg and Kj. Rasmussen
Carbohydr. Res. 69 (1979) 27-38

(6) S. Melberg and Kj. Rasmussen
Carbohydr. Res. 71 (1979) 25-34

(7) S. Melberg and Kj. Rasmussen
J. Mol. Struct. 57 (1979) 215-239

(8) S. Melberg and Kj. Rasmussen
Carbohydr. Res 78 (1980) 215-224

(9) Kj. Rasmussen
Acta Chem. Scand. A40 149-153(1986)

(10) Kj. Rasmussen
Molecular Structure and Dynamics, (M. Balaban, Ed.), Balaban, Jerusalem (1980) 171-210

(11) E. Clementi
Determination of Liquid Water Structure, Coordination Numbers for Ions and Solvation for Biological Molecules. Lecture Notes in Chemistry, Vol. 2, Springer - Verlag (1976)

(12) S. Melberg, Kj. Rasmussen, R. Scordamaglia and C. Tosi
Carbohydr. Res. 76 (1979) 23-37

(13) S.R. Niketić and Kj. Rasmussen
Acta Chem. Scand. A32 (1978) 391-400

(14) S.R. Niketić, Kj. Rasmussen, F. Woldbye and S. Lifson
Acta Chem. Scand. A30 (1976) 485-497

(15) N.C.P. Hald and Kj. Rasmussen
Acta Chem. Scand. A32 (1978) 879-886

(16) N.C.P. Hald and Kj. Rasmussen
Acta Chem. Scand. A32 (1978) 753-756

(17) S.R. Niketić and Kj. Rasmussen
Acta Chem. Scand. A35 (1981) 623-633

(18) S.R. Niketić and Kj. Rasmussen
Acta Chem. Scand. A35 (1981) 213-218

(19) Kj. Rasmussen and F. Woldbye
Coordination Chemistry-20 (D. Banerjea, Ed.), Pergamon Press, Oxford (1980), p. 219-227

(20) Kj. Rasmussen and C. Tosi
J. Mol. Struct. 121 (1985) 233-246

(21) A. Almenningen, O. Bastiansen, B.N. Cyvin, S. Cyvin, L. Fernholt and C. Rømming
Acta Chem. Scand. A38 (1984) 31-39

(22) L.-O. Pietilä and Kj. Rasmussen
J. Comput. Chem. 5 (1984) 252-260

(23) D.E. Williams
Acta Cryst. A27 (1971) 452-455

(24) D.E. Williams
Top. Curr. Phys. 26 (1981) 3-40

THEORETICAL STUDIES OF THE ENERGETICS AND DYNAMICS OF THE AQUEOUS AND IONIC ENVIRONMENT ABOUT PROTEINS: CRYSTALS OF STREPTOMYCES GRISEUS PROTEASE A

F. AVBELJ[1], J. MOULT[2], D. H. KITSON[1], M. N. G. JAMES[3] and A. T. HAGLER[1]

[1]The Agouron Institute, 505 Coast Blvd. So., La Jolla, CA 92037; [2]Center for Advanced Research in Biotechnology, National Bureau of Standards, Bldg. 221, Room A353, Gaithersburg, MD 20899; [3]The University of Alberta, Medical Sciences Building, 113th St. & 87th, Edmonton, Alberta T6G 2H7.

Many biological macromolecules, including proteins and nucleic acids, exist and operate within an aqueous environment which plays a crucial role both in determining the structural properties of the macromolecule and in its function. An understanding of the nature of the interactions between these molecules and water is therefore important in developing a full picture of the behaviour of macromolecules in biological systems. Furthermore, with the recent development of practical methods to "engineer" changes into the structure of proteins, it has become almost commonplace to synthesise new mutant proteins, with the goal of changing the specificity or catalytic efficiency of enzymes, or the optimal conditions (temperature, pH etc.) under which these enzymes operate by making simple changes in the amino acid sequence of the protein. This protein "engineering" work has given a new impetus to efforts to elucidate the principles which underly the structure and function of macromolecules and, again, an understanding of interactions with solvent constitutes an important part of these efforts. The solvent in biological systems, however, defies description at the molecular level by conventional experimental techniques due to its disordered or semi-ordered nature. To study this solvent one can turn to simulation techniques, including molecular dynamics and Monte Carlo. We have used these techniques to study the energetics and dynamics of both the water and counter ions in the crystal of the protein *Streptomyces griseus* Protease A (SGPA) (whose structure has been solved to 1.5Å resolution with an R-factor of 12.1%).

Starting with a system consisting of 2 protein molecules, 1473 water molecules, 16 sodium ions and 26 dihydrogen phosphate ions, and using periodic boundary conditions to reproduce the crystal environment, we generated 10,000,000 configurations of the solvent using the Metropolis Monte Carlo algorithm, and, starting from the 6 millionth configuration, we ran an 11 ps molecular dynamics simulation in which the proteins were kept fixed, followed by a 60 ps simulation in which both the proteins and the solvent molecules were allowed to move.

Analysis of the results of these simulations has focused on two areas. The first involves an evaluation of the ability of the methods and force field which we use to reproduce the experimentally observed structural and dynamic properties of the system. Comparison of the mean water positions from the Monte Carlo simulation with the experimental positions shows that 68 simulated waters are within 0.5Å of an observed water position. The time-averaged structures for the 2 simulated proteins over the time period 17-60 ps of the dynamics show that the RMS deviations from the experimental structure are 1.88Å and 1.46Å (for all heavy atoms). If atoms with experimental temperature factors greater than 20Å^2, which may suffer from significant systematic errors in their coordinates, are omitted from the comparison, the RMS values drop to 1.67Å and 1.25Å. We have also calculated the RMS deviations for segments of the proteins (α helices, β loops etc.) and for different residue types and are investigating possible reasons for different RMS deviations of different parts of the protein.

The second major focus of this study has been on the behaviour of the ions. Preliminary results indicate that translation of some of the ions occurs through a jump diffusion mechanism. At the beginning of the simulation, one of the phosphate counter ions underwent large oscillations in a well-defined volume of space. A large translation then took place in which the ion moved approximately 3Å in 3 ps. It then oscillated around this new position for about 10 ps. The same process of a rapid jump followed by a period of oscillation was repeated twice more during the simulation. This jump behavior of the ion can be correlated with decreases in its electrostatic energy as it moves to areas of the crystal lattice where the electrostatic potential generated by the protein is more favorable. We have also observed unexpected clusters of ions of like charge in the simulated structure. These clusters can be accounted for by examining the electrostatic environment of the ions and by taking into account ion-water interactions. This effect has led to a reinterpretation of some of the original X-ray data for this structure.

Bond lengths in transition structures and intermediates of cycloaddition reactions.

Fernando Bernardi[a], Andrea Bottoni[a], Joseph J.W. McDouall[b], Massimo Olivucci[a], Michael A. Robb[b], Glauco Tonachini[c] and Alessandro Venturini[a].

(a) Dipartimento di Chimica "G.Ciamician", Universita' di Bologna, Bologna (Italy)

(b) Department of Chemistry, King's College London, Strand, London WC2R2LS (U.K.)

(c) Istituto di Chimica Organica, Universita' di Torino, Torino (Italy).

Abstract

In this paper we have considered the following series of prototype cycloaddition reactions:
[2+2] cycloadditions $H_2C=CH_2 + H_2C=CH_2$, $H_2C=O + H_2C=O$, $H_2C=CH_2 + O=O$: 1,3 dipolar cycloadditions $HCNO + HC\equiv CH$, $HCNO + H_2C=CH_2$, $H_2CNHO + H_2C=CH_2$; [4+2] cycloaddition $H_2C=CH_2 + H_2C=CH-CH=CH_2$: and we have focused our attention upon the geometrical features of the various critical points located on the corresponding reaction surfaces. In particular we have analyzed the new forming C-C and C-O bonds, and we have derived standard values appropriate for the various types of transition structures and intermediates.

1. Introduction.

Cycloadditions are a very important class of reactions, which can be used to obtain compounds of various ring sizes[1]. Although these reactions have been largely investigated experimentally,[1,2] considerable controversy still surrounds their mechanism.[1c-d,3]

A cycloaddition reaction generally involves the formation of two new σ bonds between the reactants at the expense of π bonds. For such processes it is possible to postulate three different mechanisms:

i) a synchronous concerted approach involving a cyclic transition state (TS) with the two new bonds formed to equal extent:

ii) a two-stage asynchronous concerted mechanism in which there are two distinct stages to changes in bonding, some occurring mainly between the reactants and the single TS and the others mainly between the TS and products.

iii) a two-step process, which occurs in two kinetically distinct steps via a stable diradical intermediate.

Recently we have computed the complete potential energy surfaces for the following series of prototype cycloaddition reactions:[4]

i) the [2+2] cycloadditions $H_2C=CH_2 + H_2C=CH_2$[4a], $H_2C=O + H_2C=O$,[4b] $H_2C=CH_2 + O=O$:[4c]

ii) the 1,3 dipolar cycloaddition $HCNO + HC \equiv CH$, $HCNO + H_2C=CH_2$, $H_2CNHO + H_2C=CH_2$.[4d]

iii) the [4+2] cycloaddition $H_2C=CH-CH=CH_2 + H_2C=CH_2$.[4e]

The various potential energy surfaces have been computed with ab-initio MC-SCF techniques using minimal (STO-3G)[5] and extended (4-31G)[6] basis sets. All critical points have been fully optimized using MC-SCF gradient techniques[7] and characterized by diagonalizing the related Hessian matrices computed using finite differences. Integral and derivative calculations have been performed using the Gaussian 80 series of programs,[8] while the CI and MC-SCF codes used are described in ref.9. In all cases we have used a CAS (Complete Active Space) wavefunction. The valence space, that we have chosen, contains four orbitals in all cases except for the reaction ethylene-singlet oxygen where we have used a valence space with six orbitals. For technical reasons we have used a valence space with six orbitals also in the 4-31G study of the reaction between fulminic acid and ethylene.[4d]

In previous papers we have mainly discussed the mechanistic aspects associated with these surfaces, i.e. the nature of the various critical points, the possible reaction paths and the relative energy differences. In this paper we focus our attention upon the geometrical features of the various critical points and in particular upon the values of the forming bonds. The critical points of interest are first and second order saddle points (that for sake of simplicity we refer to as transiton structures) and diradicaloid intermediates. The purpose here is to perform a comparative analysis of the computed bond lengths and examine if there are trends in the various cases in order to derive standard values for these bonds which allow chemists to construct with greater accuracy the geometries of transition structures and intermediates.[10] The computations performed so far allow us to discuss only the following bonds: C-C and C-O.

2. Values of the bond lengths

In order to perform meaningful comparison it is convenient to distinguish the following situations:

i) bonds in cyclic transition structures, where the two new bonds are formed to equal extent;

ii) bonds in diradicaloid transition structures where only one new bond is formed;

iii) bonds in diradicaloid intermediates.

2.1 Values of the C-C bonds

Typical values of the C-C bond for various situations are reported in Figure 1 together with a schematic representation of the corresponding molecular structures.

We must point out that the structures shown in Figure 1 do not correspond to all the structures that we have located investigating the various potential energy surfaces. In various cases there exist in fact other diradicaloid structures and intermediates which have almost coincident C-C bond lengths and which, for this reason, have not been included.

For instance, in the case of the [2+2] reactions, in addition to the trans structures shown in Figure 1, we have also located gauche structures which differ just in the value of the CCCC dihedral angle. Similar situations have been found for the other reactions.

These computational results allow to compare either covalent situations, arising from the interaction of covalent species (as it is the case for the ethylene+ethylene and butadiene+ethylene situations), or polar situations, where at least one of the interacting species has a polar character as it is the case for the 1,3 dipolar cycloadditions and the [2+2] cycloadditions involving formaldehyde.

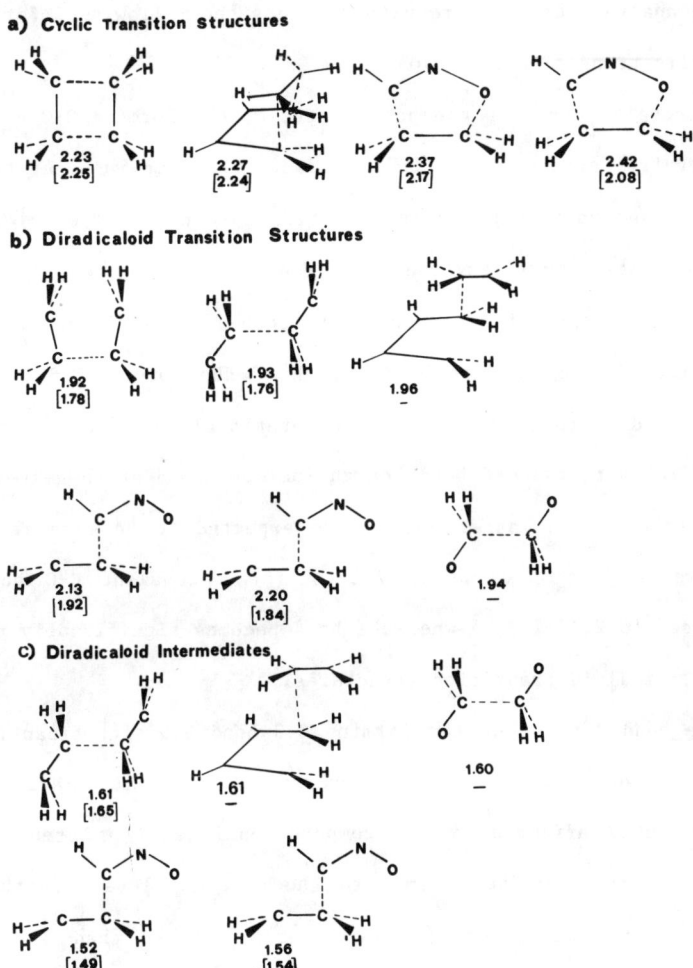

Figure 1. Optimized values of the forming C-C bonds in the various transition structures and intermediates computed at the STO-3G and 4-31G (brackets) computational level.

The analysis of these results provides the following informations
i) Cyclic transition structures.

In covalent species the bond-length of the forming C-C bond appears to be fairly constant. We find, in fact, the same value at the minimal and extended basis set and in a forbidden (ethylene + ethylene) and an allowed (butadiene + ethylene) reaction.

In the presence of polar reactants such as fulminic acid, there is a significant change in the bond length of the forming C-C bond and its trend is different at the two computational levels. At the minimal level, in fact, the C-C bond length increases and at the extended level it decreases. The latter values are expected to be more reliable and thus we propose a value of 2.24 Å for a covalent C-C bond, which decreases to 2.17-2.08 Å when the bond becomes significantly polar.

ii) Diradicaloid transition structures.

The bond lenghts of the forming C-C bond are significantly shorter than in cyclic transition structures. These values are also more significantly affected by the computational level and tend to become shorter going from the minimal to the extended level. Furthermore in the covalent situations the values are shorter than in the polar situations. In this case we propose a value of 1.78 Å for a covalent C-C bond that increases to 1.84-1.92 Å when the bond becomes significantly polar.

iii) Diradicaloid intermediates.

In these cases the values of the formed C-C bond are almost not affected by the basis set level and an increase of bond polarity determines a slight shortening of the bond-length. For these species we propose a value of 1.60 Å for a covalent C-C bond that decreases to 1.55-1.50 Å when the bond becomes polar.

2.2 Values of the C-O bonds

The values of the C-O bonds in the various typical available situations are reported in Figure 2. The results associated with the ethylene-singlet oxygen cycloaddition represents a covalent situation since the interacting fragments are covalent species, while those associated with the formaldehyde-formaldehyde and the 1,3 dipolar cycloadditions represent polar situations since the interacting fragments are already polar species. It can be seen that in all cases these values are not significantly affected by the basis set level. In the cyclic transition states the values tend to increase with the increase of polarity, while the opposite trend is found in the diradicaloid transition states. On the other hand in the diradicaloid intermediates the values in the different situations are almost coincident.

We propose the following sets of values:
i) in the cyclic transition states, for the C-O bond in covalent situations, a value of 2.10 Å which becomes 2.30 in polar situations.
ii) in the diradicaloid transition states, for the C-O bond in covalent situations, a value of 1.85 Å which becomes 1.65 Å in polar situations.
iii) in the diradicaloid intermediates, a value of 1.53 Å.

a) Cyclic T.S.

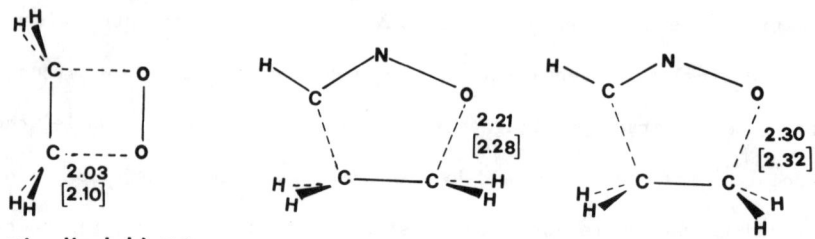

b) Diradicaloid T.S.

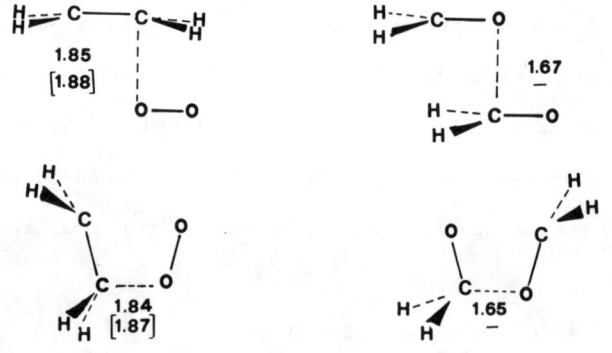

c) Diradicaloid Intermediates

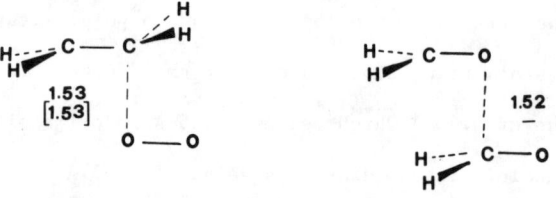

Figure 2. Optimized values of the forming C-O bonds in the various transition structures and intermediates computed at the STO-3G and 4-31G (brackets) computational level.

3. Conclusions

In this paper we have presented a comparative analysis of the geometrical features of various critical points located on the potential energy surfaces of cycloaddition reactions. In particular we have examined in detail the new forming C–C and C–O bonds in various types of transition structures and intermediates and for different situations (covalent and polar) of the reactant molecules. We have also investigated the dependence of the optimum values of these bonds on the accuracy of the basis set which has been used (minimal or extended). This type of analysis has allowed us to define structural reference values for the C–C and C–O bond lengths in the various situations that we have examined.

The reference values that we have proposed for the C–C and C–O bonds can be used by organic chemists as a basis for developing a flexible model for constructing reliable geometries of transition structures and intermediates in the case of other cycloaddition reactions.[10]

In developing such a model we also need a method for defining reliable values of the intra-fragment geometrical parameters. A simple but rather accurate recipe is to take, either for transition structures or intermediates, the average between the values in the reactants and the products. A theoretical justification of this recipe is given by the diabatic surface model, which shows that the transition structures originate from the crossing between a diabatic surface associated with the bonding situation of the reactants and a diabatic surface associated with the bonding situation of the products.[11] Furthermore the MC-SCF results have shown that the geometries of the diradical intermediates are very similar to those of the corresponding fragmentation transition states except for the values of the forming bonds.

A further possible and more appealing use of the proposed bond length values can be found in the development of a quantitative treatment by force-field (molecular mechanics)^{12} models of the reactivity properties of cicloaddition reactions and possibly, in the future, of other classes of reactions.

The nature of the force-field method, in fact, is such that it is suitable for dealing with molecular systems for which experimental data (like geometries, torsional energies, force constant) can be available. Problems immediately occur when we try to describe systems which do not lie at energy minima like transition states or in general systems which have very short lifetimes, like intermediates, for which we cannot apply the most common experimental methods for the study of structures and energies.

The MC-SCF results can be used to provide the appropriate parameters (geometrical parameters, force constants...) which are necessary for developing a force-field treatment of transition structures and intermediates in cycloaddition reactions.

References:

1. (a) R.Huisgen, Angew. Chem. Int. Edit. Eng., 2, 565 (1963); (b) R.Huisgen, R.Grashey and J.Sauer in "The Chemistry of Alkenes", S.Patai Ed. Interscience, London, 1964, pp.741-953; (c) J.Sauer, Angew. Chem. Int. Edit. Eng., 5, 211 (1966): (d) J.Sauer, ibid., 6, 16 (1967)
2. (a) R.B.Woodward and T.J.Katz, Tetrahedron, 5, 70 (1958); (b) J.G.Martin and R.K.Hill, Chem. Rev., 61, 537 (1961); (c) S.B.Needleman and M.C.C. Kuo, 62, 405 (1962): (d) J.D.Roberts and C.M.Sharts, Org.

React. <u>12</u>, 1, (1962)· (e) R.Huisgen, Angew. Chem. Inter. Ed. Engl. <u>7</u>, 321 (1968): (f) H.Kwart and Kenneth King, Chem. Rev., <u>68</u>, 415 (1968): (g) R.Gompper, Angew.Chem.Inter.Ed.Engl., <u>8</u>, 312 (1969)· (h) J.Sauer and R.Sustmann, ibid., <u>19</u>, 779 (1980).

3. (a) R.A.Firestone, J.Org.Chem., <u>33</u>, 2285 (1968) (b) W.C.Herndon, Chem.Rev. <u>72</u>, 157 (1972): (c) K.N.Houk, Joyner Sims, C.R.Watts and L.J.Luskus, J.Am.Chem.Soc., <u>95</u>, 7301 (1973): (d) K.N.Houk Acc.Chem.Res., <u>8</u>, 361 (1975): (e) R.Huisgen, J.Org.Chem., <u>41</u>, 403 (1976): (f) R.A.Firestone, Tetrahedron, <u>33</u>, 3009 (1977): (g) R.B.Woodward, R.Hoffmann, Angew. Chem. Inter. Ed. Engl., <u>8</u>, 781 (1969)

4. (a) F.Bernardi, A.Bottoni, M.A.Robb, H.B.Schlegel and G.Tonachini, J. Am. Chem. Soc., <u>107</u>, 2260 (1985): (b) F.Bernardi, M.Olivucci, M.A.Robb, in preparation (c) G.Tonachini, H.B.Schlegel, F.Bernardi and M.A.Robb, J.Mol.Structure THEOCHEM., <u>138</u>, 221 (1986): (d) J.J.W.McDouall, M.A.Robb, V.Niazi, F.Bernardi and H.B.Schlegel, J. Am. Chem. Soc., in press: (e) F.Bernardi, A.Bottoni, M.A.Robb, M.J.Field, I.H.Hillier and M.F.Guest, J.Chem.Soc.Chem.Commun., 1052 (1985): F.Bernardi, A.Bottoni, A.Venturini, M.A.Robb, M.J.Field, I.H.Hillier and M.F.Guest, J.Am.Chem.Soc., in press.

5. W.J.Hehre, R.F.Stewart and J.A.Pople, J.Chem.Phys., <u>51</u>, 2657 (1969)

6. R.Ditchfield, W.J.Hehre and J.A.Pople, J.Chem.Phys., <u>54</u>, 724 (1971)

7. H.B.Schlegel and M.A.Robb, Chem.Phys.Lett., <u>93</u>, 43 (1982)

8. J.S.Binkley, R.A.Whiteside, R.Krishan, R.Seeger, D.J.Defrees, H.B.Schlegel, S.Topiol, L.R.Kahn and J.A.Pople, QCPE, 13, 406 (1981)

9. R.M.A.Eade and M.A.Robb, Chem.Phys.Lett., <u>83</u>, 362 (1981)

10. C.A.Reynolds and C.Thomson, J.Chem.Soc., Faraday Trans. 2, <u>83</u>, 961 (1987)

11. (a) F.Bernardi and M.A.Robb, Mol.Phys., <u>48</u>, 1345 (1983); (b)

F.Bernardi and M.A.Robb J.Am.Chem.Soc., <u>105</u>, 54 (1984); (c) F.Bernardi, S.Paleolog, J.J.McDouall and M.A.Robb, J.Mol.Struct. THEOCHEM., <u>138</u>, 23 (1986)

12. (a) N.L.Allinger, M.T.Tribble, M.A.Miller and D.H.Wertz, J.Am.Chem.Soc. <u>93</u>, 1637 (1971); (b) N.L.Allinger in "Adv. in Phys.Org.Chem.", <u>13</u>, 1 (1976): (c) M.R.Imam and N.L.Allinger, J.Mol.Struct., <u>126</u>, 345 (1985); (d) D.C.Spellmeyer and K.N.Houk, J.Am.Chem.Soc., <u>52</u>, 959 (1986).

SOME ASPECTS OF COMPUTATIONAL POLYMER QUANTUM CHEMISTRY

J.M.André
Facultés Universitaires ND de la Paix
Laboratoire de Chimie Théorique Appliquée
61, rue de Bruxelles
5000 Namur
Belgium

1. INTRODUCTION: THE ROLE OF POLYMERS IN MODERN LIFE.

In a paper devoted to computational polymer quantum chemistry, it is worth to first develop the role of macromolecules and their impact on modern technology. Then, the main concepts of a quantum mechanical description of linear chains and quasi-1 dimensional systems will be introduced. This will provide the opportunity of discussing specific topics of such a 1-dimensional solid-state physics. As applications, some aspects of interest in relation with XPS spectroscopy experiments on polymers will be reviewed. We shall also pay attention to some aspects of bond alternation in 1-dimensional chains in connection with the newly discovered metallic properties of doped conjugated polymers, a field which became widely studied in the last ten years. The last example briefly reviews a tentative molecular design of the optoelectrical properties of organic polymers. The underlying interest in these selected examples is that theory has turned out to be a guide and a motivation for further experiments.

Figure 1 is a molecular model of a polymeric chain. It is an idealized representation of polyethylene. Polyethylene is widely used as plastic bags and films ; on the chemical point of view, it is a member of the alkane series, C_nH_{2n+2}: methane, ethane, propane, butane, ...In Figure 1, polyethylene is viewed as a backbone of carbon atoms in tetrahedral sp^3 situations.

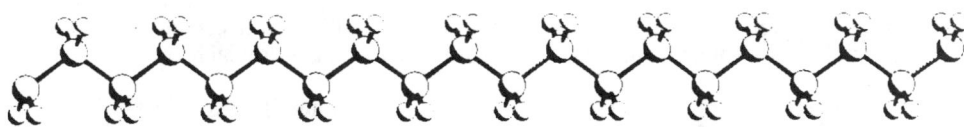

Figure 1. Molecular representation of an oligomer of polyethylene, C_xH_{2x}

The concept of macromolecules is recent; it was introduced in 1921 by Herman STAUDINGER (1881-1965), the father of modern polymer chemistry (Nobel Prize in 1953) but has been rather difficult to admit. An historical criticism against Staudinger has been largely reported in the literature[1,2]: "Dear Colleague, leave the concept of large molecules well alone; organic molecules with a molecular weight above 5000 do not exist. Purify your products, such as rubber, then they will crystallize and prove to be lower molecular substances. Organic molecules with more than 40 carbon atoms do not exist. Molecules cannot be larger than the crystallographic unit cell, so there can be no such things as a macromolecule". This comment is only sixty years old. It was not until the middle of the 30's, that Staudinger's ideas were fully recognized. Nowadays, not only the existence of macromolecules is well established but also their values in modern technology. It has been recently evaluated [3] that 25% of synthetic polymers are used in packaging, 21% in building and construction industry, 15% for electrical and electronic purposes, 10% as glues, cements, paints, coatings, 7% in car industry, 5% for home furnitures and furnishings, 2.5% as cooking articles. On an average, a car contained 10 kg of plastics in 1960, 17 kg in 1966, 48 kg in 1972, 60 kg in 1979, and 85 kg in 1980. Let us mention also the whole field of biopolymers.

On a volume basis, the US annual production of plastics exceeds that of copper and aluminum and crossed that of steel during the beginning of the 80's [4]. As at the same time, the trends in plastics price over the last years have been consistently downwards in contrast to the trends for most other materials, the plastics business has become far more competitive over the last few years. Table 1 [5] details the US annual production of the main artificial polymers over the last decade.

Table 1: Production by the U.S. Chemical Industry (Millions of lb).

	1976	1980	1984	1985	1986
Man-made fibers	8130	9547	9486	8122	8447
Synthetic rubbers	2303	2015	2155	1838	1987
Plastics					
Thermosetting resins	3633	4093	5549	5631	5848
Thermoplastics resins	21810	26621	33744	35202	37745

If the oil crisis has evidently lowered the production increase during the last years, 30 billion pounds produced in the United States are nevertheless the enormous quantity of about 150 pounds per year and per person. Even in the life products, we are concerned with very important amounts of biopolymers. Each man in the world has about 750 grams of hemoglobin; multiplied by 4.6 billions of individuals, this makes 3 millions tons of hemoglobin. If polymers are made from very simple elementary blocks : oil, hydrogen, oxygen, chlorine, nitrogen and sulfur, the resulting products are in some cases of very high social value; let us think of synthetic arteries and artificial hearts.

Polymers have raised up fascinating discoveries in scientific research. An example is that of SN_x (polysulfurnitride) whose the assumed electrical behavior is listed in table 2 as a function of the years. It was synthesized around the beginning of this century and characterized as an explosive powder without any peculiar electrical property. Papers until 1960's considered it as a good insulator. In 1964, evidence was given that SN_x was a semi-conductor; in 1973, the proof was given that it is a metal. Finally in 1975, it was shown at IBM San Jose to become a superconductor at a very low temperature. Other examples concerning doped conjugated polymers will be developed later on.

Table 2: SN_x (polysulfurnitride).

1910	Insulator	A new sulphide of nitrogen F.P.Burt, J.Chem.Soc., 1121, **1910**
		The sulfur nitrides $(SN)_2$ and $(SN)_x$ M.Goehring & D.Voigt Naturwissenschaften, **40**, 482, **1953**
1964	Semi-conductor	Spectra and the **semi-conductivity** of the SN_x polymer D.Chapman et al. Trans.Far.Soc., **60,** 294,**1964**
1973	Metal	Polysulfur nitride- a one-dimensional chain with a **metallic** ground state V.V.Walatka, M.M.Labes & J.H.Perlstein Phys.Rev.Letters, **31**, 1139, **1973**
1975	Superconductor	**Superconductivity** in polysulfur nitride SN_x R.L.Greene, G.B.Street & L.J.Suter Phys.Rev.Letters, **34**, 557, **1975**

In view of the existence of quantum physics, of quantum chemistry, of quantum biology, and of quantum pharmacology, on one side, and of the role of polymers in the everyday life and of their interesting properties, on the other side, our opinion was, fifteen years ago, that there was a timely need for developing a specific quantum polymer chemistry. It was the main concern of the research pursued in our laboratory and summarized in the next paragraphs of this paper.

2. BAND STRUCTURE THEORY OF POLYMERS: MOLECULAR ORBITAL VERSUS SOLID STATE CONCEPTS.

In the second part of this paper, we present some facets of the standard quantum mechanical treatment of polymers. Pioneering quantum mechanical calculations on a polyethylene chain were made around the end of the 60's[6,7]. In polymer quantum chemistry, polyethylene plays the role of the helium atom in atomic physics and of the hydrogen molecule in molecular physics. Polyethylene is a zigzag chain built from methylene groups. The limitations sketched in Figure 2 are implicitly contained in our standard elementary calculations.

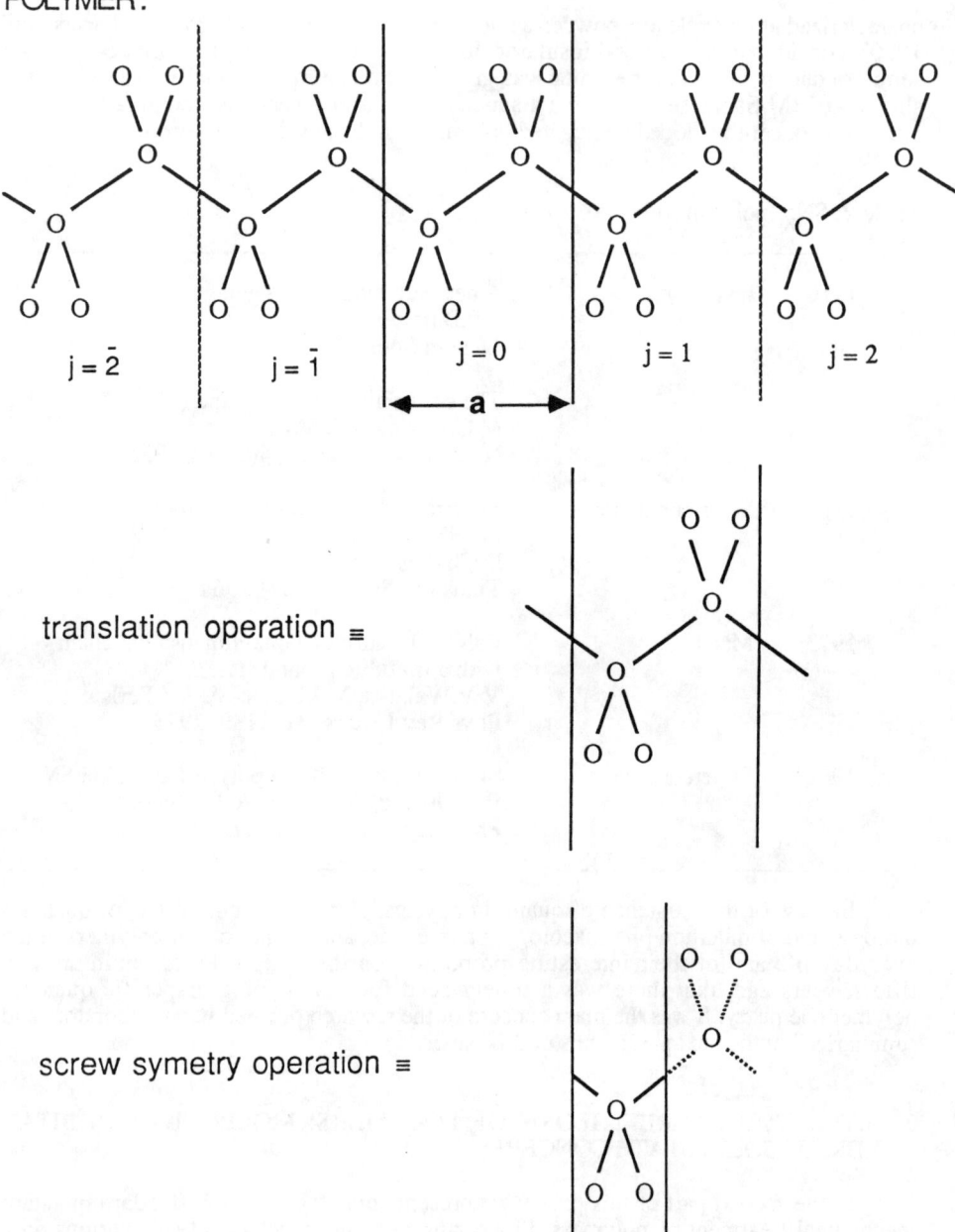

Figure 2. Model and unit cells of a polyethylene chain

First, we consider an isolated chain; in practice, the chain is never isolated but does exist in a liquid or solid state environment. Secondly, the chain is considered as infinite and periodic. The fact of considering the chain as infinite is not a strong limitation since polymers can have very large molecular weights; in the case of usual polymers like polyethylene or polypropylene, one observes molecular weights in the range 100.000 up to 10.000.000. The model also neglects the end chain effects; the importance of end effects has not yet been studied in detail. The strongest assumption is that of the translational symmetry of the linear chain which allows to use the language of solid-physicists in order to get a rather complete description of the electronic structure of polymers.

In molecules and polymers, the standard theory is based on the Hartree-Fock theory. In this independent model, a single electron moves in the field of the nuclei and in the mean Coulombic and exchange field of all the other electrons. A set of molecular orbitals (MO's) is obtained to describe the occupied and unoccupied one-electron wave functions. In molecular quantum chemistry, the molecular orbitals are drawn as single levels which are at most doubly occupied by a pair of electrons of opposite spins. On the other hand, the solid state physicist takes advantage of the possible translation symmetry of the lattice and uses the concept of Brillouin zones introduced by Bloch in 1928. In this theory, the so called Bloch functions (molecular orbitals for an infinite 1D chain) are eigenfunctions of a translation operator. Bloch's theorem is a direct consequence of the periodicity of the electron density:

$$|\phi_n(r)|^2 = |\phi_n(r+ja)|^2$$

where a is the length of the polymeric unit cell. By taking the square root in the complex mathematical space, Bloch's theorem states the phase relation of the orbitals at periodically related points:

$$\phi_n(r+ja) = e^{ikja} \phi_n(r)$$

Since the argument of an exponential is a pure number, a is a length, j is a pure number, the counter index of a given unit cell, k must have the dimensions of an inverse length. Thus, the orbitals and their associated energies are functions of that k and are labelled by it:

$$\phi_n = \phi_n(k)$$
$$\varepsilon_n = \varepsilon_n(k)$$

They can be plotted with respect to k; the representation of the corresponding dispersion curves is called an energy band. The energy bands are periodic in k-space so that our knowledge of the full k-energy dependence is reduced to a single unit cell of the reciprocal space:

$$\varepsilon_n(k+lg) = \varepsilon_n(k)$$

The search is simplified if we use a symmetrized part of the reciprocal space with respect to k=0, the so called first Brillouin zone ranging from $-\pi/a$ to $+\pi/a$. Due to further symmetries:

$$\varepsilon_n(-k) = \varepsilon_n(k)$$

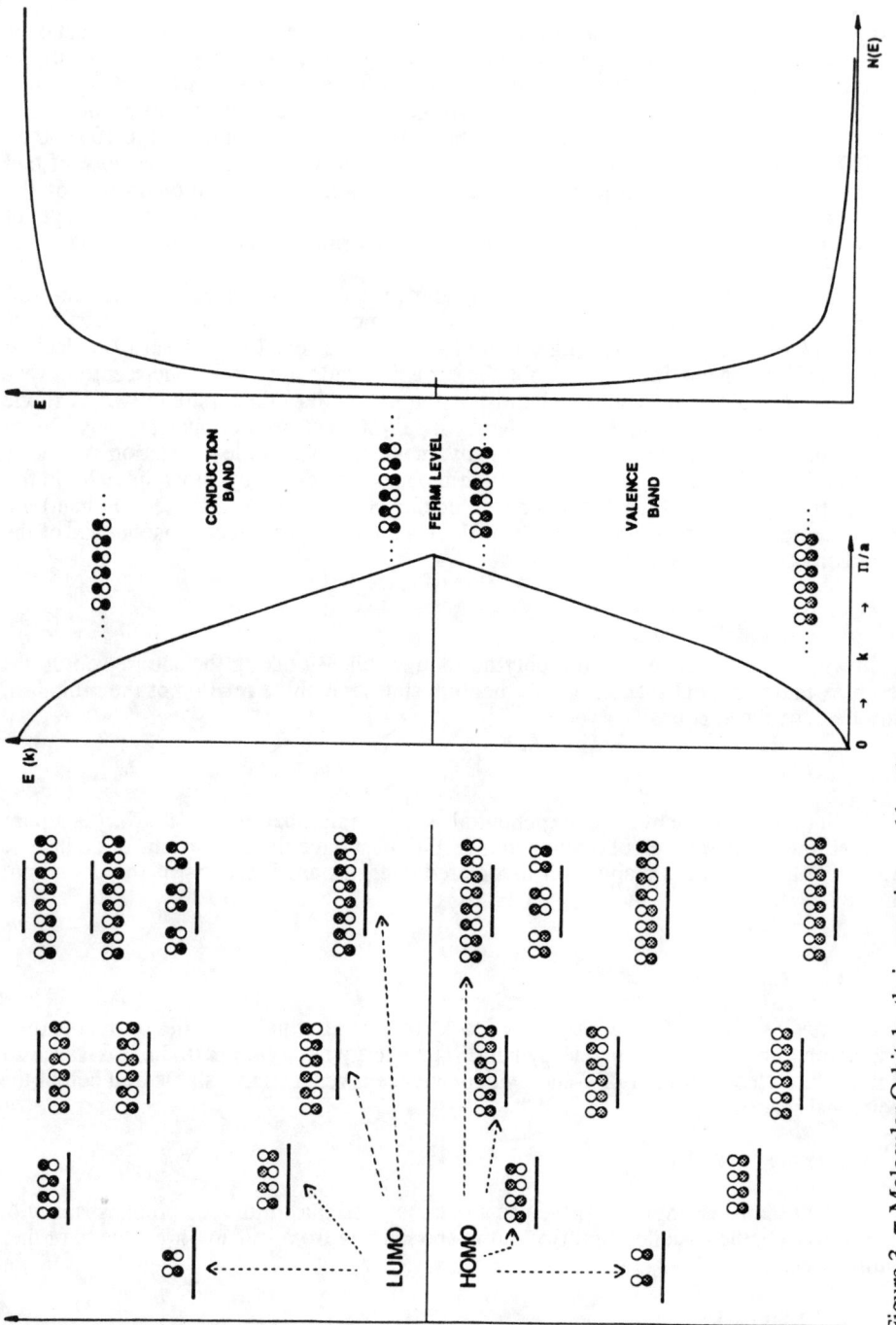

Figure 3. π-Molecular Orbitals, their energies, and band structure of polyacetylene

the reciprocal space must be only explored from k=0 to k=+π/a (half the first Brillouin zone).

Figure 3 states the links between solid state concepts (which are not too familiar to chemists) and molecular theories (which are not too familiar to physicists). If he is interested in evaluating the electronic properties of a polymer or of a large oligomer, the molecular quantum chemist will probably start by an extrapolation study as the one summarized in Figure 3. That figure is an adaptation, to the case of polyacetylenes, of a classical figure explaining the properties of planar graphite given in the pioneering book by the Pullmans on "quantum biochemistry"[8]. It has been frequently used since for the polyethylene (methane, ethane, propane,...) series [9,10] and even previously for the polyene-polyacetylene series (ethylene, butadiene, hexatriene,...) [11-13] in the framework of the Hückel method. In its leftmost part, Figure 3 plots the MO levels as a function of the number of carbon atoms in the oligomeric chain. It can be noticed that as the number of carbon atoms increases the distance between the energy levels diminishes in order to form, in the infinite limit, occupied energy regions (the internal and valence bands), unoccupied energy regions (the conduction bands) and forbidden energy regions (the forbidden bands or energy gaps). In the selected example of the center part of Figure 3 (polyacetylene), the translation unit cell contains a -CH=CH- unit and thus 2 π-electrons. The first Brillouin zone contains the corresponding 2 valence bands.

The rightmost representation of Figure 3 is the easiest one for experimental comparisons; it is the density-of-states (DOS) which plots the number of available energy levels as a function of energy for infinite systems. It is usually normalized to the number of electrons per unit cell. The three representations are equivalent. When expanding the MO's in terms of atomic orbitals (AO) by the LCAO approximation, the number of obtained MO's is equal to the size of the AO basis. Due to first principles, the number of nodes increases with orbital energy. On the left part of Figure 3, the bonding and antibonding π orbitals of ethylene are easily recognized as are also the four orbitals of butadiene popularized by the work of Woodward and Hoffman [14]. In band theory, the polymeric orbitals are real and can be plotted easily at high symmetry points of the first Brillouin zone. The lowest orbital has no node (symmetric combination of AO's, $\Sigma \exp(ikja) \chi_j = \Sigma 1 \chi_j$ if k=0) while the highest has the maximum number of nodes (most antisymmetric combination, $\Sigma \exp(ikja) \chi_j = \Sigma(-1)^j \chi_j$ if k=π/a). Intermediate situations occur at both sides of the Fermi level for the Highest Occupied Molecular Orbital (HOMO) and the Lowest Unoccupied Molecular Orbital (LUMO). At other points of the first Brillouin zone, the polymeric orbitals are complex but still obey the same principles.

Figure 3 appears to be a crossing link between molecular quantum chemistry and solid state physics and, as a short comment, polymer quantum chemistry turns out to be an excellent field to force chemists to become familiar with basic concepts of solid state physics. In Europe, there still unfortunately exists a too strong separation between the academic curricula of chemists and physicists.

Figure 4 exemplifies the case of an hypothetical chain built from unit cells containing a single π orbital. The symmetric (bonding) combination of the orbitals is obtained at k=0. The orbital energy increases and the energy band has a positive derivative towards the point k=π/a where the antisymmetric (most antibonding) combination is observed.

In Figure 5, we consider an unit cell containing two orbitals. It is the case of an all-trans regular polyene. The Bloch functions are thus constructed from the well-known

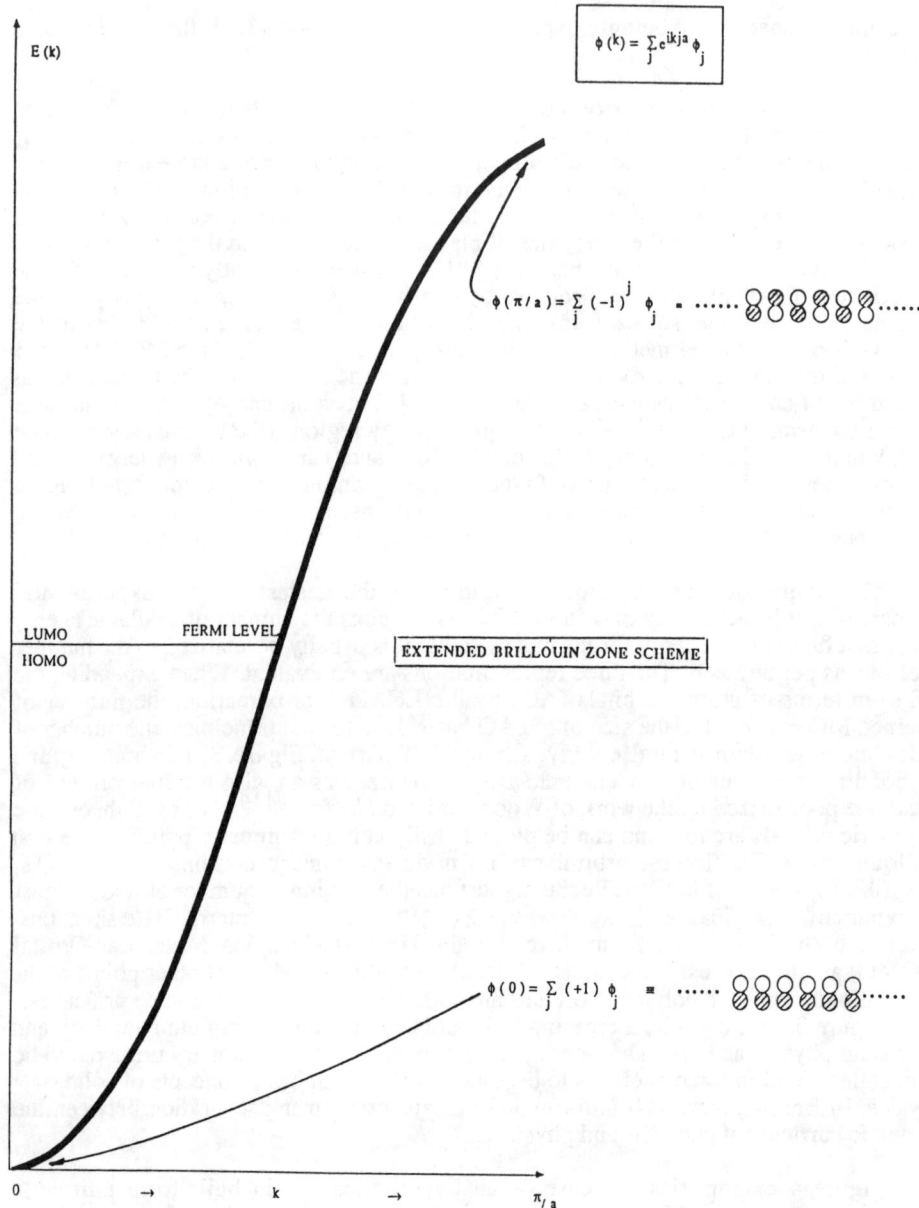

Figure 4. Qualitative band structure of a chain with a single-π-orbital per unit cell

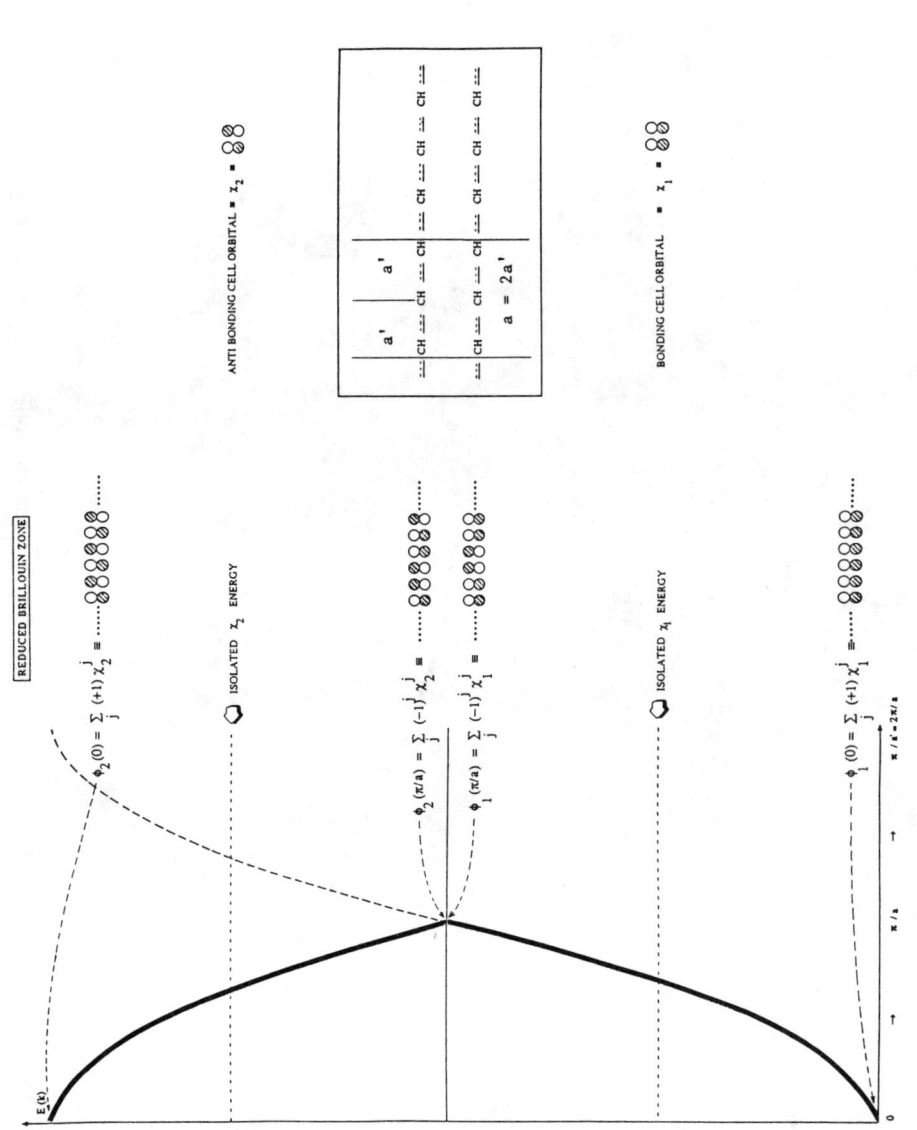

Figure 5. Qualitative band structure of a chain with two-π-orbitals per unit cell

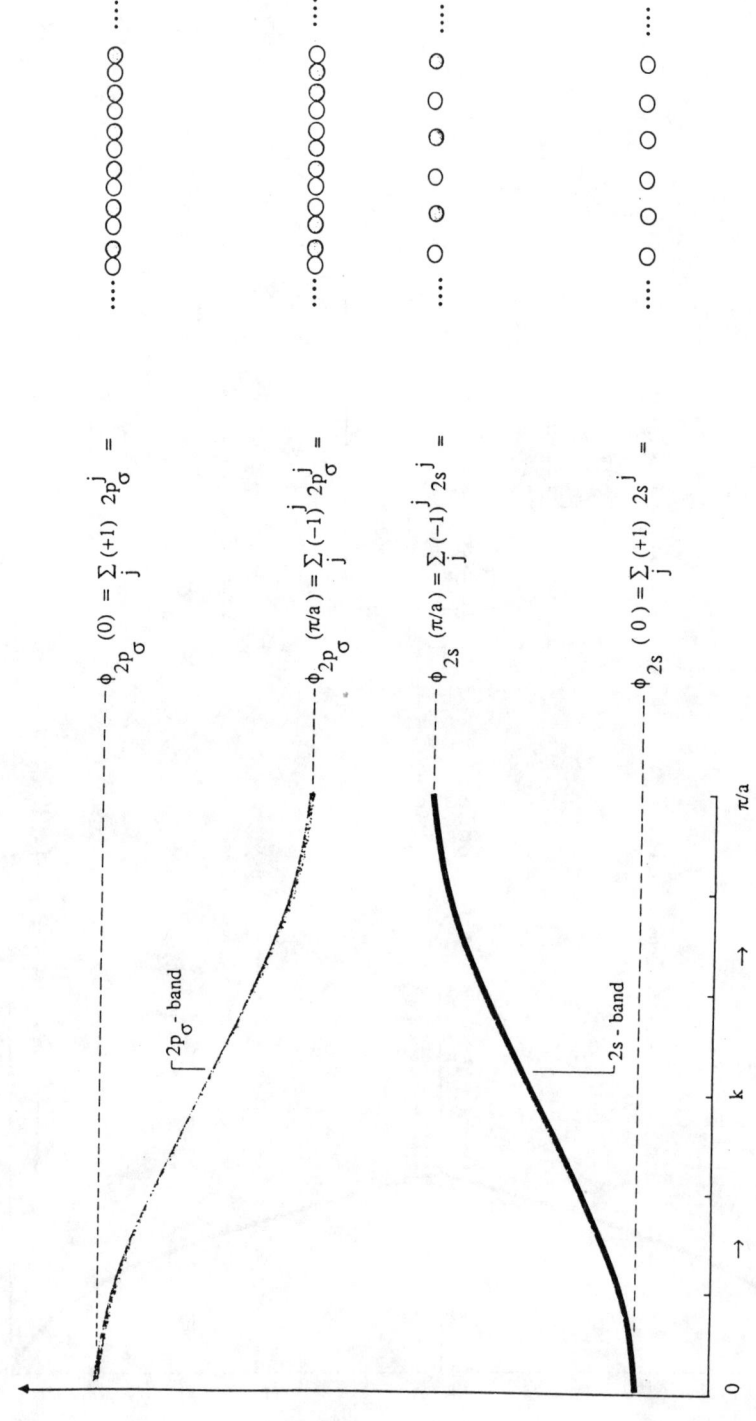

Figure 6. Qualitative band structure of an hypothetical chain with one s- and one π-orbital per unit cell

π-bonding orbital and π-antibonding orbital of ethylene. With respect to the previous case (a single orbital and a single energy band), two orbitals are centered in the unit cell and two energy bands appear in the first Brillouin zone. The Brillouin zone is half its original size since the unit cell is twice larger. At k=0, we observe the positive combination of both basis orbitals and the antisymmetric combination at the edge of the zone (k=π/a). The lower band has a positive derivative (increasing number of nodes from the center to the edge of the zone); the higher band has a negative derivative (decreasing number of nodes from the center to the edge of the zone). The sign of the derivative of the energy band gives thus an indication on the bonding and antibonding characters of the various orbitals within the first Brillouin zone.

Figure 6 shows the orbital combination of s and pσ bands. At k=0, the positive combination of the orbitals means the most bonding situation for the s orbitals and the most antibonding one for the pσ. At k=π/a, the opposite situation is observed, i.e., the most antibonding situation for the s orbitals and the most bonding for the pσ ones. The s-band has a positive derivative while the pσ-one has a negative one.

Such interpretations are of basic interest when investigating the bonding and antibonding behavior of the orbitals for designing highly conducting polymers as will be mentioned in part 4 of this paper. Qualitative molecular orbital techniques based on the previous principles are now commonly used to sketch π-band structures of conjugated polymers [15,16]

An important comment is that polymer quantum chemistry is not a one-dimensional solid state physics. In the strictly 1D physics, the systems are periodic in one dimension and have 1D wave-functions. In polymer quantum chemistry, the systems and their wave-functions are three-dimensional but periodic only in one direction. Usual theorems of 1D physics are consequently no longer valid. A typical example is that of extrema of the energy bands which should only occur at the center and the edges of the Brillouin zone in a strictly 1D system. For polymers, even in simple cases like polyethylene, some extrema of the energy bands are encountered at arbitrary positions of the first Brillouin zone.

Turning to actual applications, the numerical procedure combines the equations of the methods of molecular quantum chemistry and of solid state physics (table 3).

In molecular quantum chemistry, a molecular orbital is expanded in terms of basis functions; secular systems of equations and determinants are solved; their eigenvalues are the orbital energies. From the LCAO coefficients, charges and bond orders (projection of the density matrices onto the limited basis used) are calculated. In polymer quantum chemistry, we take into account the lattice periodicity; the orbitals, the systems of equations and the determinants are no longer real but have imaginary components. This introduces lattice sums which are to be evaluated by adequate procedures[17-19]. The key problem is to obtain the matrix-elements over the basis functions. Fortunately, these matrix-elements exponentially decrease with the distance between the orbital centers and force the natural convergency properties of the lattice sums. They are calculated by the standard methods of quantum chemistry cited in table 4.

Table 3: A comparison of some basic formulas of molecular and polymer quantum chemistry.

Molecular Quantum Chemistry	Polymer Quantum Chemistry
$\phi_j(\mathbf{r}) = \sum_p c_{jp} \chi_p(\mathbf{r}-\mathbf{P})$	$\phi_n(k,\mathbf{r}) = N^{-1/2} \sum_h e^{ikha} \sum_p c_{np}(k) \chi_p(\mathbf{r}-\mathbf{P}-h\mathbf{a})$
$\sum_p c_{jp} (F_{pq} - \varepsilon_j S_{pq}) = 0$	$\sum_p c_{np}(k) \{ \sum_h e^{ikha} [F_{pq}^h - \varepsilon_n(k) S_{pq}^h] \}$
	$\equiv \sum_p c_{np}(k) \{ F_{pq}(k) - \varepsilon_n(k) S_{pq}(k) \} = 0$
$|(F_{pq} - \varepsilon S_{pq})| = 0$	$|\{ \sum_h e^{ikha} [F_{pq}^h - \varepsilon(k) S_{pq}^h]\}|$
	$\equiv |\{ F_{pq}(k) - \varepsilon_n(k) S_{pq}(k)\}| = 0$
$\rightarrow \varepsilon_1, \varepsilon_2, \varepsilon_3, ...$	$\rightarrow \varepsilon_1(k), \varepsilon_2(k), \varepsilon_3(k), ...$
$\rightarrow D_{pq} = \sum_j c_{jp} c_{jq}$	$\rightarrow D_{pq}^h = a/2\pi \sum_n \int dk\, c_{np}^*(k) c_{nq}(k) e^{ikha}$

Table 4: The empirical, semi-empirical, and ab initio methods of quantum chemistry.

	non self-consistent	self-consistent
π electrons	Hückel Theory (HMO) Hückel	ω-technique Pariser-Parr-Pople technique Pariser-Parr Pople
valence $\sigma + \pi$ electrons	Extended Hückel (EHT) Hoffmann LCLO, SAMO Duke Leroy Valence Effective Hamiltonian VEH Durand & Nicolas	Zero Differential Overlap CNDO INDO MINDO MNDO Pople Dewar
all electrons	Floating Spherical Gaussian Orbital Frost	ab initio Clementi → IBMOL Pople → GAUSSIAN Whitten → Gaussian Lobes

The methods are classified as self-consistent and non self-consistent methodologies. In the self-consistent methods, the effect of electron-electron interactions is explicitly taken into account. In the ab initio methods, all necessary integrals are correctly evaluated up to infinity as soon as the geometry and the basis functions are defined. A correct computation is still a formidable task at the present time. Indeed, it has been proved that long range effects are very important as soon as the unit cell contains permanent dipoles. Multipole expansion must be used in order to get a satisfactory balance of the electrostatics interactions. Less sophisticated ab initio methods are available and produce results of good quality in much less computer time. It has been sometimes advocated to use cheaper semi-empirical techniques like Extended Hückel or CNDO; in these cases, reliable results are obtained when paying a very special attention to the parametrization procedures. First sketches of the conjugated bands are easily obtained from simple Hückel calculations. The techniques which try to simulate the ab initio results are also of particular interest. Good results have been recently obtained with simple model potential techniques (the so called VEH-Valence Effective Hamiltonian technique). In this procedure, the Hartree-Fock operator is approximated by a sum of atomic potentials which are, in general, anisotropic projectors determined on model molecules.

The practice of polymer computations is rather standard. The Fock and overlap matrices are computed at a given level of approximation (semi-empirical or ab initio). The calculation is made in direct space. The k-dependent matrices are diagonalized in the reciprocal space and, if necessary, a self-consistent procedure is turned on. The output consists of the energy bands and of the form of the molecular orbitals of the polymer. An interactive graphical communication can be initiated for ordering the energy bands, plotting the standard electronic properties, as band structures, band widths, density of states, simulations of electron spectra, electron densities, calculating electron indexes as charges and bond orders, and determining conformations or other properties. Automatic programs taking into account the effects of long range interactions are fully implemented.

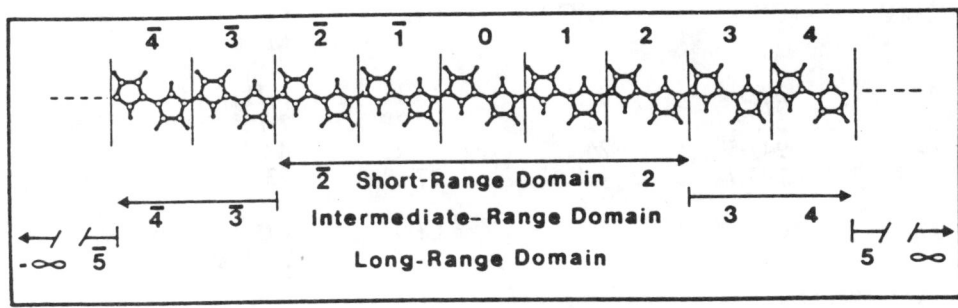

Figure 7. Model for an ab initio computation on polypyrrole

For example, a computation of the electronic structure of polypyrrole has been performed on a full ab initio scale[20]. The unit cell taken into account for the polymer calculations contains two pyrrole units, i.e., $N_2C_8H_6$. All two-electron integrals are evaluated for a region covering nine unit cells (18 pyrrole units) as depicted in Figure 7. In the so called short-range domain, extending over five unit cells, the two-electron integrals are exactly calculated. In the intermediate-range domain, covering four more unit cells (two on each side of the short-range region), the two-electron integrals are evaluated with a numerically precise asymptotic form of the F_v functions which are involved in calculations with Gaussian basis sets. A multipole expansion technique[21] including all dipole-dipole and monopole-quadrupole interactions (k+l+1=3) is used beyond the intermediate region up to infinity, in order to take account the long-range Coulombic contributions.

Most of the computer programs imply the translational symmetry but also allow for helical symmetry. For example, polyethylene can be studied with a unit cell of two CH_2's or, alternatively, with a unit cell containing a single CH_2 which is translated by a screw axis of order 2 [22] (see, for example, Figure 2). This can perhaps be thought to be of not too much importance but it can produce enormous savings of computer time in actual calculations; Teflon, an usual synthetic polymers, where all the hydrogens of polyethylene are replaced by fluorines, is an helix of order 17 while polyalanine, a natural biopolymer, has a screw axis of order 47.

Our computational work has been made on several types of computers. Hückel band structures were obtained as early as 1965 on an IBM 1620[23]. The first ab initio band structure of polyethylene[7] was computed on an IBM 360/91. Lot of the FSGO band structure calculations[24] have been made on a PDP 11/45. The polypyrrole band structure previously cited[20] was obtained on an IBM 4341. Presently, we use the Namur Scientific Computing Facility (noted as SCF for obvious reasons to the quantum scientists).

Table 5: Diagonalization in band structure calculations (CPU time in seconds). (IBM= IBM 4341/2, FPS=FPS 164

Matrix Size	IBM	IBM +FPS	IBM +FPS	IBM +FPS	
Opt 0	CBORIS	CBORIS	CH	CH MTRANS	Ratio
19	2.50	0.71	1.01	0.21	11.9
57	65.90	17.17	25.49	2.76	23.9
64	94.05	24.05	36.40	4.07	23.1
126	703.72	178.17	273.28	23.74	29.6
Opt 3	CBORIS	CBORIS	CH	CH MTRANS	Ratio
19	0.90	0.14	0.16	0.16	5.6
57	22.81	3.25	3.02	2.31	9.9
64	33.05	4.95	4.75	3.49	9.5
126	243.52	32.82	32.57	21.52	11.3

19 = Polyacrylonitrile
57 = Helical polyacrylonitrile
64 = Polyparaphenylene oxide
126 = Polyimid

As schematized in Figure 8, that configuration consists of two IBM 4341 and one attached processor FPS 164. This is a joint experimental project between the university of Namur, the National Foundation for Scientific Research (FNRS-NFWO) and IBM-Belgium. Table 5 presents in some cases relevant to the present paper the significant gain of computing time observed on the configuration. It compares several CPU times in the basic complex non-orthogonal diagonalization (HC=SCE) procedure for several size of basis sets (19,57,64,126). The table indicates, for two levels of the compiler (Opt 0 and Opt 3), the CPU times obtained by using a standard FORTRAN diagonalization procedure (CBORIS) or a specially coded routine (CH/MTRANS) for the attached processor FPS 164. Calculations have been made on the IBM 4341 alone and on the IBM 4341 + FPS 164 configuration. With respect to the stand-alone configuration, speed gain attains a factor of about 25. The result is that typical batch jobs for the calculation of band structures of polymers now become almost real time jobs; they can, at the limit, be performed interactively and thus allow for a true theoretical molecular design of polymers with specific properties.

In the first part of this paper, it has been shown that polymer quantum chemistry is a new science which links the concepts of quantum chemistry with those of solid state physics. It has become an interdisciplinary and timely concern of our scientific community.

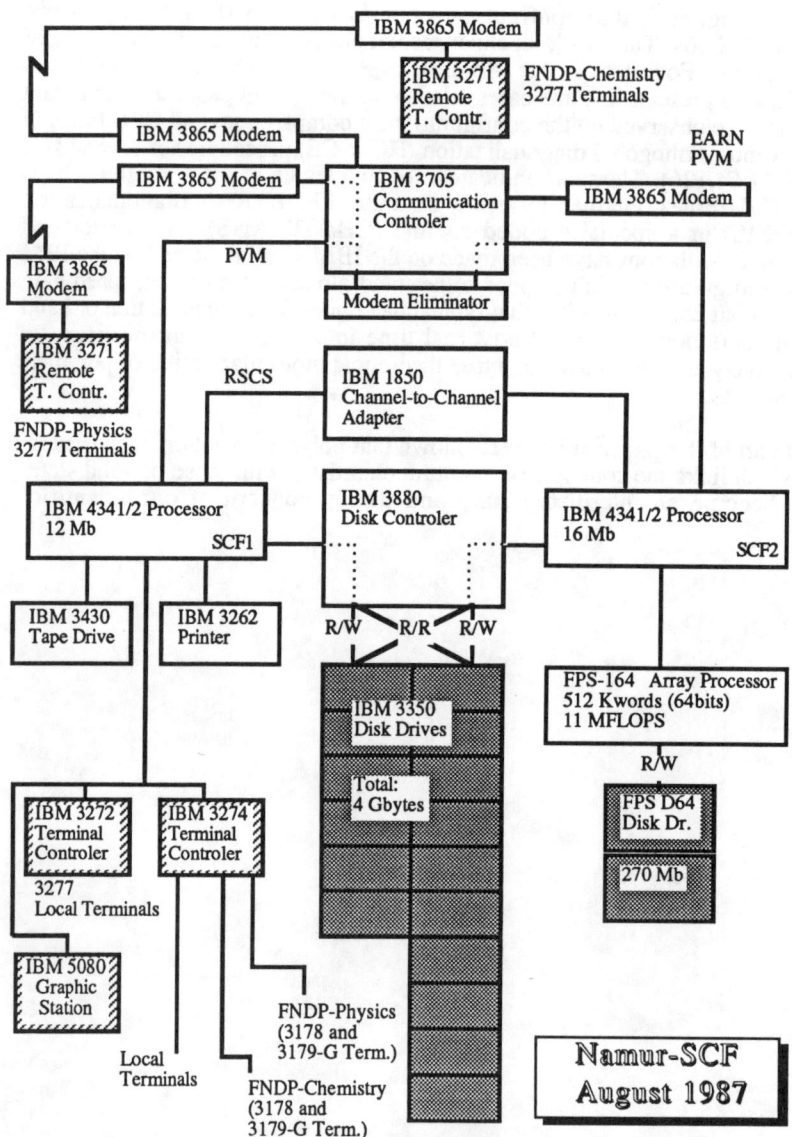

Figure 8. Namur-SCF configuration

3. A FIRST EXAMPLE OF APPLICATION: XPS INTERPRETATIONS OF POLYMERIC PROPERTIES.

From the beginning of the 70's, applications of quantum mechanical methods to polymers became increasingly important. One of the several reasons for that is the improvement of experimental techniques for investigating electronic properties of polymers like photoelectron spectroscopy (PS) and especially X-ray induced photoelectron spectroscopy (XPS), sometimes known as ESCA. The development of the later has been mainly due to the efforts of Kai Siegbahn (Nobel prize in 1981) in Uppsala. In polymer chemistry, two pioneering papers were published in 1972. They refer to the experimental ESCA analysis of the core levels of fluoropolyethylenes[25] and of the valence band of polyethylene[26]. Theoretical interpretations rapidly followed to support the interpretation of the ESCA core levels[27]. The theoretical analysis of ESCA valence spectra has been less direct. The ESCA spectrum and a semi-empirical band structure of polyethylene [26] are sketched in Figure 9. It shows that band structure is not measurable directly and one has to apply transformations to bring calculated data in a form readily comparable to experiment.

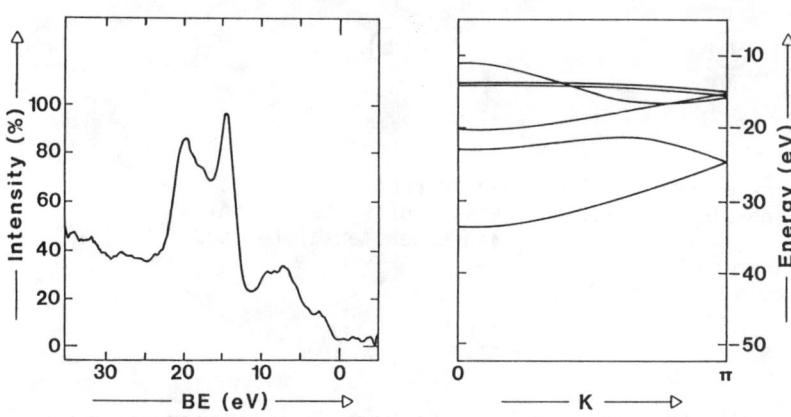

Figure 9. Sketch of the theoretical band structure and experimental density of states of polyethylene (after Ref.26)

In author's opinion, band structure plots do not offer the best representation of valence bands properties particularly in those systems where a large number of bands lies in a narrow energetic region as it is the case for polymers. This is a reason why we have implemented and systematized a three step procedure [28] which leads to a theoretical simulation of an XPS valence spectrum of a regular polymer as illustrated in figure 10.

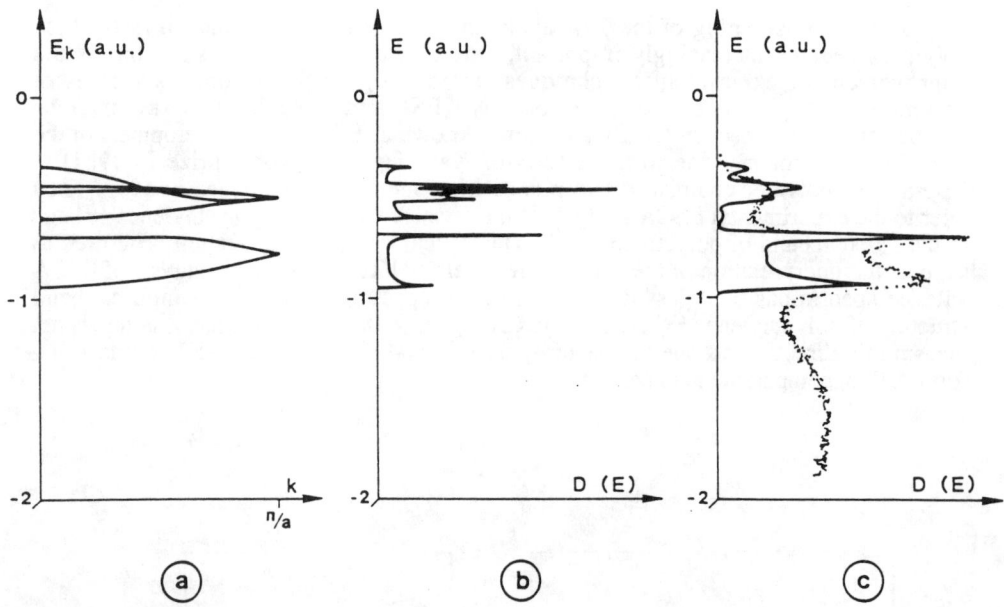

a) One-electron valence band structure of polyethylene,
b) one-electron valence density of states of polyethylene,
c) simulated (full line) and experimental (dotted lines) XPS valence spectrum of polyethylene.
Calculations in a FSGO model

Figure 10. FSGO Theoretical simulation of valence photoelectron spectra of polyethylene

In its presently simplified form, this procedure corresponds to (a) the calculation of the density of states (DOS) histograms, (b) taking into account cross-section effects, and (c) convolution for simulating experimental resolution. In Figure 10, this procedure has been applied to the test case of the FSGO valence bands of an all-trans polyethylene chain [29]. The positions of both theoretical and experimental peaks agree surprisingly well and both fine structures are directly comparable. From the bond order analysis of the theoretical calculations, the four highest bands can be labeled as contributing to the C-H bonds while the two lowests correspond to the C-C bonds. Figure 10 is also basically important since it states the existence of an experimental basis to the concept of electron bands in polymers.

A natural development of this work was to study the effects of structural perturbations like conformational changes on the valence bands of stereoisomers. To illustrate the approach, a computer experiment has been made in the case of polyethylene [30]. A vibrational analysis [31] of crystalline samples suggests the possibility of four conformations of polyethylene: trans (T), gauche (G), trans-gauche (TG) and trans-gauche-trans-gauche (TGTG'). The theoretical calculations reveal important differences in the shape of the densities of states of those four conformers of polyethylene as illustrated in Figure 11. The conformational changes do affect the C-C bands.

Their bottom energy values are almost constant while a modification takes place at the top of the bands giving rise to important changes in the band widths. Another aspect is that an energy gap appears in the C-C band for which there is successive trans-gauche conformations (TG and TGTG'). The T and G forms are easily distinguished by their fine structure.

Unfortunately, the corresponding samples are not available and this computer experiment only supports the presumption that conformational effects could be experimentally observable by photoelectron spectroscopy measurements of valence bands. It is incentive to investigate a real case and prove the existence of observable conformational effects on the valence electronic DOS. A tentative interpretation of the experimental ESCA spectrum of isotactic polypropylene has been reported [32] assuming a fully extended zig-zag chain conformation in the calculation. No satisfactory agreement as to the peak structure was obtained. The problem has been reinvestigated by explicitly considering the DOS of polypropylene in its actual isoclined 2*3/1 helical form. As a

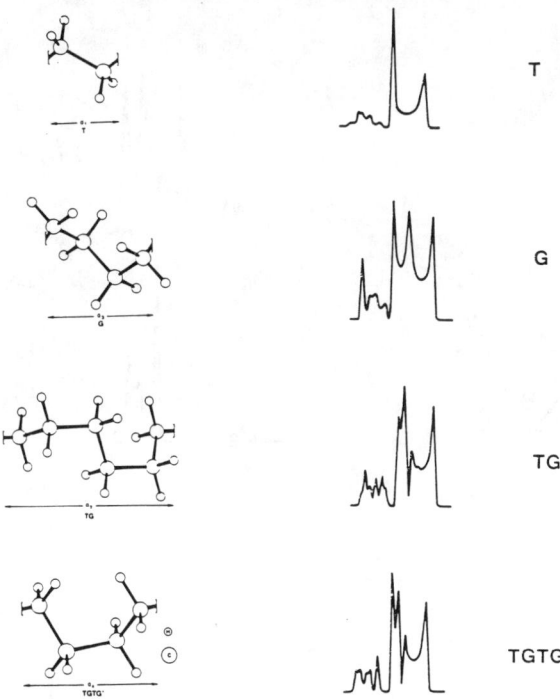

Figure 11. Four conformations of polyethylene and their simulated VEH photoelectron spectra

consequence, the right structure appears in the theoretical spectrum [30] as illustrated in Figure 12. The conformational effects have also been studied in the case of syndiotactic polypropylenes [22] which exist in a zig-zag and in a two-order helical form.

These few examples demonstrate the mutual enrichment which has been gained from a close interrelation between theory and experiment in the field of photoelectron spectroscopy.

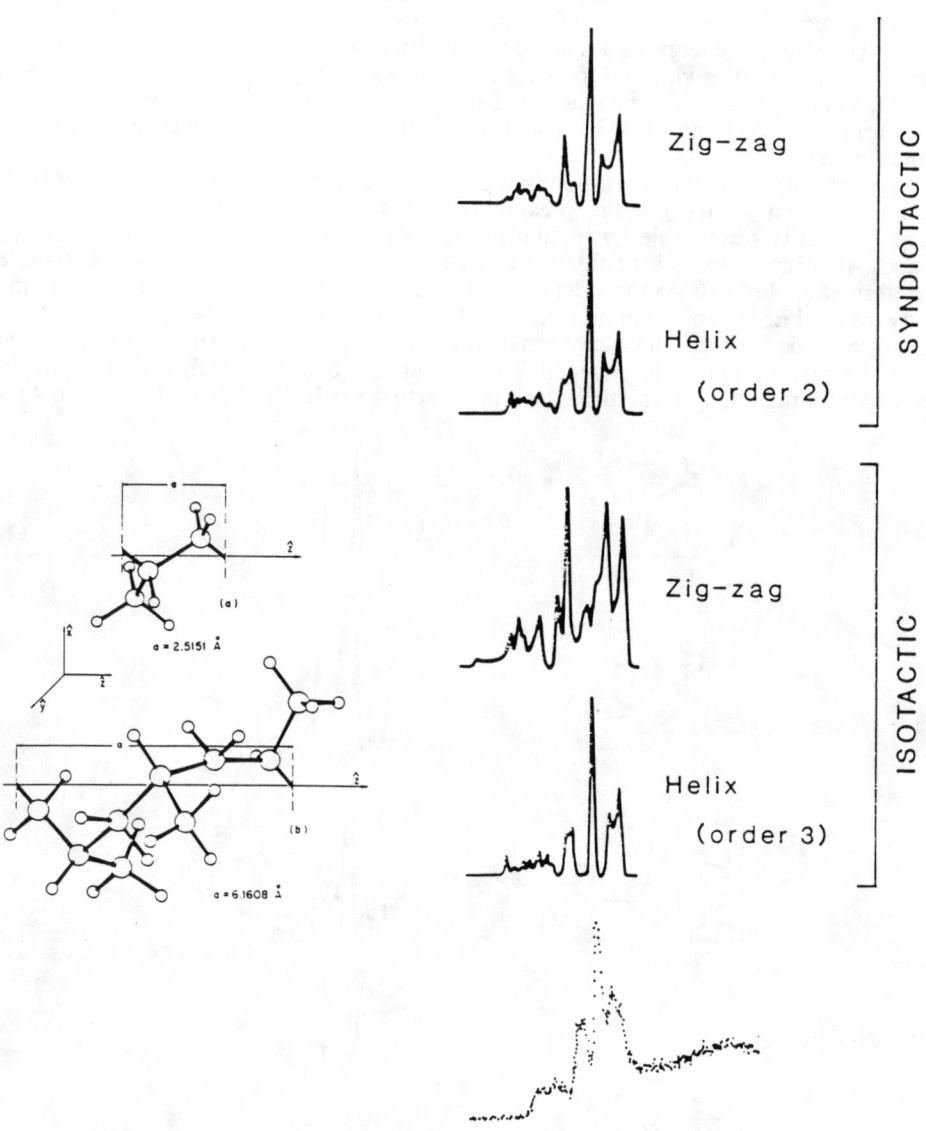

Figure 12.   Theoretical (VEH) and experimental (when available) spectra of isotactic zig-zag and helical polypropylene

4. A SECOND EXAMPLE OF APPLICATION: A STUDY OF THE ELECTRONIC STRUCTURE OF SOME HIGHLY CONDUCTING POLYMERS.

In recent years, the discovery of doped organic polymers with high conductivity has generated substantial research interest among physicists and chemists alike. On one hand, the creation of new materials combining the processability, light weight and durability of plastics with the electrical conductivity of metals is a driving force in the development of conducting polymers. On the other hand, doped organic polymers constitute a new fascinating area of condensed-matter physics where non-linear phenomena can play an important role. Polymers with doped derivatives are reported to have conductivities larger than 1 S/cm and they include conjugated systems such as polyacetylene, polyparaphenylene, polythiophene and polypyrrole. The doping process involves exposure of the polymer to electron donors (such as alkali metals) or acceptors (such as I_2 or AsF_5). Although doped polymers display phenomena in some way similar to conventional doped inorganic semiconductors, their physics is very different. One of the fundamental difference is that these polymers are organic materials. Therefore, it is expected that charge-transfer (or electron excitation) processes will result in significant local modifications (relaxations).

The conductivities can be varied in such systems from twelve to sixteen orders of magnitude. The standard idea on conductivity is that a metal should have a zero energy gap. Depending on the size of the energy gap, a semiconducting or insulating state is observed. In the case of organic polymers, the existence or the non-existence of energy gaps was already related to the concept of bond alternation by the pioneering work of Kuhn [33-35] during the end of the 40's. He has shown that in the series of polymethine dyes, the bond lengths between all the carbon atoms are the same due to a resonance balanced between equivalent extreme forms. All carbon-carbon bonds in the skeleton have 50% of double bond character. This fact was later confirmed by X-ray diffraction studies. The electrons are completely free; there is no energy gap between the valence and conduction bands and the limit of the first UV-visible transition for an infinite chain is zero. On the opposite, when the same approach is used in the case of polyenes (polyacetylenes), no resonance forms are possible and a much poorer agreement with experiment is obtained. In order to reproduce the experimental data, Kuhn has to include a sinusoidal perturbative potential so that the electron distribution corresponds to alternating single and double bonds. He had to postulate an alternation of the bond lengths between single and double-like bonds. In this case, the extrapolation of the UV spectrum tends to a non-zero energy gap for the infinite chain.

In simple Hückel calculations, an infinite chain of sp^2 carbon atoms is in a metallic state if all the interactions are considered as equal (perfect regular chain). An energy gap opens up in the band structure if alternating bond interactions are used. This forbidden gap is in complete agreement with the ideas developed by Kuhn in the framework of a free-electron model. It is also related to Peierls' distortion: an one-dimensional metallic chain is unstable and tends to distort in order to become semiconducting or insulating. In a first approximation, conductivity means equalization of bonds. The following computer experiment schematized in figure 13 has confirmed this point[36].

An isolated chain of equidistant hydrogen atoms has strongly alternating electron density indices (the elements of the density matrix between adjacent hydrogens), a clear evidence of the geometrical tendency to bond alternation. When doped by lithium, there is

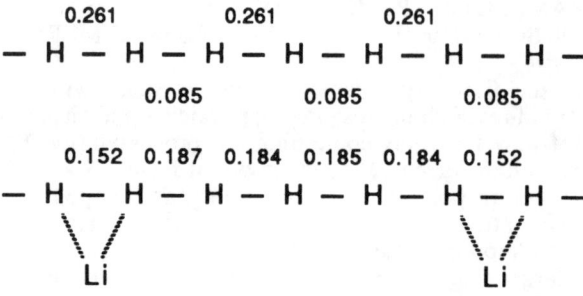

Figure 13. Effect of doping on an hypothetical chain of hydrogen atoms. STO 3G results

an important charge transfer between the chain and the dopant and the electron density is strongly modified as indicated by the changes in electron density indices. There is a net trend towards an equalization of the H-H bonds. These results indicate from a very simple computer experiment that high doping induces strong geometric modifications on the polymer chain (i.e., in this case, a tendency of bond alternation to decay). Furthermore, there exists a concentration at which the changes seem more pronounced.

A chain of hydrogen atoms is an easy model well adapted to simulate polyacetylene; each atom of the chain of hydrogen atom has a single 1s electron. Each carbon atom of the polyacetylene has a single π electron. Like in polyacetylene, the infinite chain of hydrogen atoms exhibits bond alternation. The bond alternation is even so strong that it dissociates into hydrogen molecules. More sophisticated calculations have been published [37]. A realistic model of polyacetylene has been studied. The external doping has the same effect as in the case of hydrogen chains. In a LCAO minimal STO-3G basis set, the equilibrium geometry of a single polyacetylene chain is found strongly alternating (about 1.33 Å for double bonds and 1.48 Å for single ones). In conjugated chains, STO-3G basis underestimates the lengths of double bonds and overestimate those of the single ones. An explicit determination of the position of the doping agent (lithium in the calculation) has been made. For the all-trans polyacetylene chain, lithium is found to occupy the out-of-plane center of the hexagonal positions made by four carbons and two hydrogens. The interesting effect is that the amount of bond alternation is drastically reduced, 1.41 and 1.43 Å, respectively, instead of 1.33 and 1.48 Å, previously. The same type of calculations has been performed on cis-chains.

However, it is essential to note that the energy gap is never fully removed and that other mechanisms must be involved for explaining the electrical conductivity. The idea of solitons in polyacetylene was implicitly introduced in a classical and important paper by Pople and Walmsley[38] in 1962. The soliton is a radical misfit which could exist in the middle of a long polyene chain and which consists of several successive bonds of similar lengths near which the odd electron is localized. The defect would be mobile and, if charged, this could explain the electrical conduction in charged polyenes. This idea has been largely developed in the Su, Schrieffer, and Heeger (SSH) theory [39].

Actual calculations of the soliton defects have been performed. They show the extreme importance of local geometry deformations for explaining the electrical conductivity of doped polyacetylene. The energetics of the interactions of soliton defects has been studied in detail by Brédas [40]. Polyacetylene is unique among the studied systems in possessing a degenerate ground state represented in Figure 14. Soliton theory requests the resonance forms of the polymer chain to be degenerate and is unable to explain the electrical conductivity of systems like polyparaphenylene (PPP) where, as shown in Figure 14, the quinoid form has a higher energy than the phenyle resonance form.

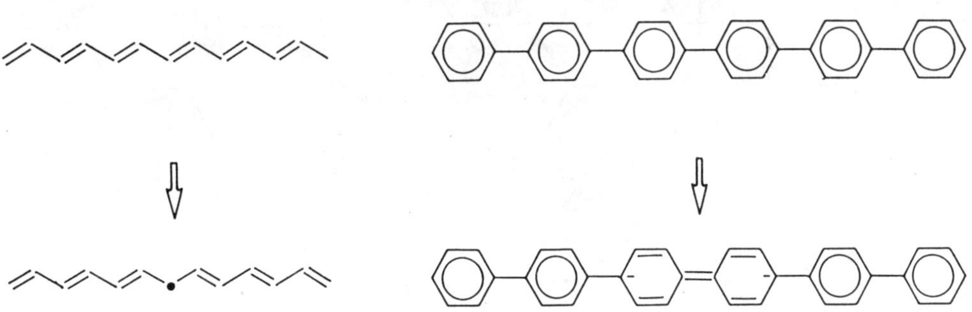

Figure 14. Resonance forms of polyacetylene and polyparaphenylene

STO-3G estimation of the energy difference between the benzenoid and quinoid forms is 20.1 kcal/mol per ring. The changes of geometry of an oligomer containing four monomers[41], the quaterphenyle oligomer, are sketched in figure 15.

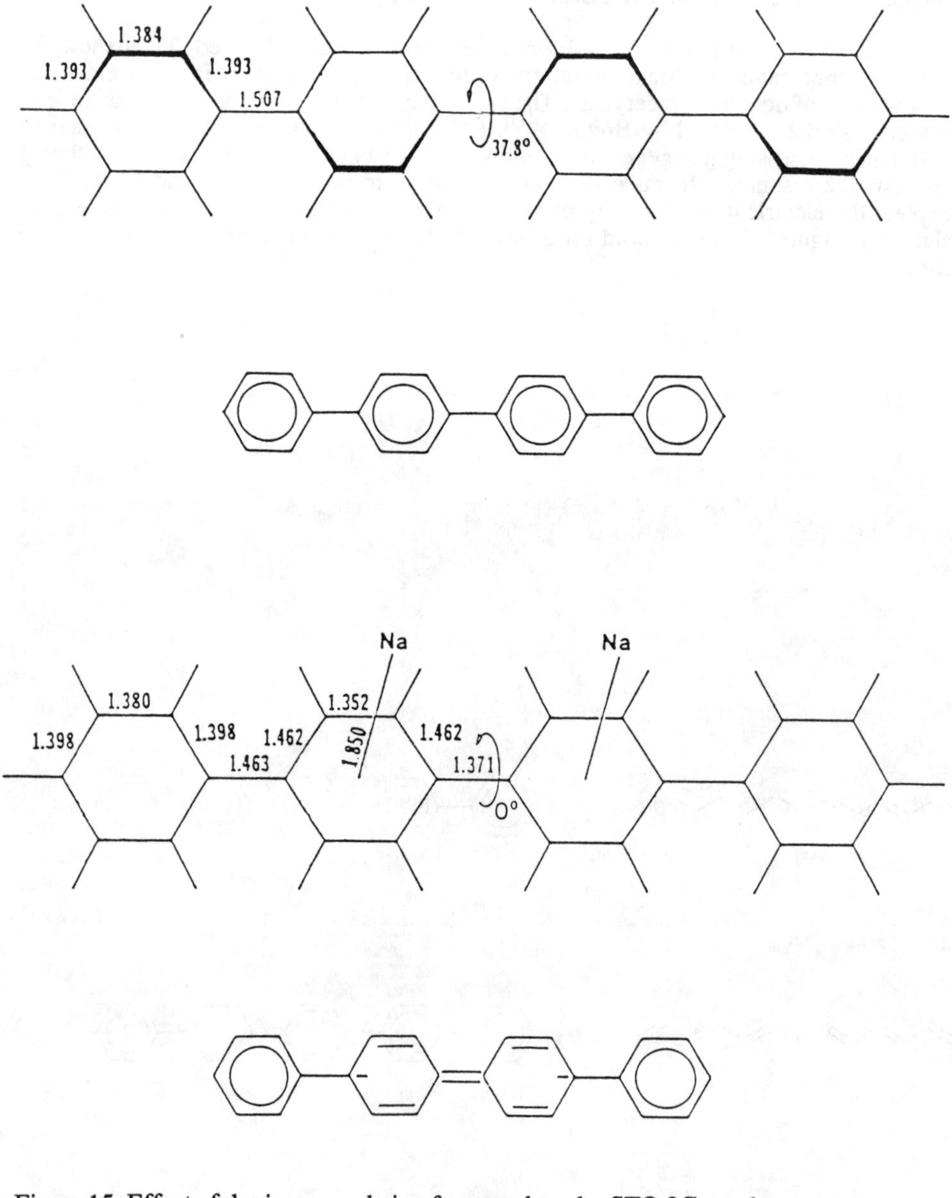

Figure 15. Effect of doping on a chain of quaterphenyle. STO 3G results

In the undoped chain, benzenoid structures are connected together through single bonds (1.51 Å). There exists a torsion angle of 38° between two subsequent phenyles. If the systems become more coplanar, the delocalization would lead to a lowering of the gap. When doping by Li, the transfer is .64 e⁻ per lithium atom and the angle is reduced to 2°. In the case of sodium doping, the charge transfer is almost complete, 0.93 e⁻ per atom and the chain is coplanar. A quinoid structure is formed. The C-C bond between the rings is markedly reduced to 1.398 Å. Parallel bonds in the rings acquire a more pronounced double-bond character as they decrease by 0.010 Å. Inclined bonds increase significantly by 0.044 Å. In the case of sodium doping, the inner rings are almost purely quinoid with 1.353 and 1.378 Å corresponding to the double bonds and 1.469 Å for the single bonds. As in the polyacetylene case, the gaps are drastically reduced without reaching however a zero value. An explanation of the metallic properties of PPP can be found in terms of polarons which induce important modifications on the electronic structure[42].

Figure 16 presents the π band structure of (a) a PPP chain considered with the STO 3G geometry of undoped quaterphenyle, (b) a polyquaterphenyle chain with the geometry of sodium-doped QP, and (c) a polyquinoid chain having the geometry of the inner quinoid rings. Situations (a), (b), and (c) can thus be viewed as corresponding, to 0, 50 and 100 % doping levels respectively. Note that, since the unit cell for the polyquaterphenyle chain is 4 times as large as that for PPP and polyquinoid chains, there are 4 times as many p bands in the middle band structure.

In the first case, we obtain the usual PPP π band structure containing six π bands. The band gap is 3.5 eV. With respect to the undoped case, the level corresponding to the highest occupied molecular orbital is strongly pushed up in energy, whereas the level corresponding to the lowest unoccupied molecular orbital is strongly pushed down in energy. All the other levels remain more or less at the same energies as in the undoped compound. The energy calculation indicates that, in comparison to the undoped molecule, the upward shift of the HOMO is 0.74 eV and the downward shift of the LUMO is 0.88 eV upon sodium doping; these numbers have a lower value (0.50 and 0.62 eV) upon lithium doping since in the latter case the geometric modifications are smaller. The two new states in the gap are due to the charge transfer and the geometry changes implied by the formation of a doubly charged bipolaron defect on the chain. These states in the gap can then be referred to as spinless bipolaron states and are fully occupied in the case of n-lithium- or sodium-doping as we consider here. They would be empty in the case of p-doping. STO-3G calculations show that the bipolaron states are dominated by the C p orbitals. The impurity states (i.e., those related to the dopant atoms) appear above the bipolaron states. The Na 3s impurity states are located within the conduction states and not within the gap.

In the second part (b) of Figure 16, as the doping level increases, bipolaron states start overlapping and this process leads to the formation of bipolaron bands within the gap at 50% doping (the maximum concentration usually achieved experimentally); the bands have widths of about 0.55 eV. Upon application of an electric field, the possible motion of the doubly charged bipolarons along the chains and between the chains leads to a conduction mechanism without spin. This conduction mechanism is very unusual since all bands are either totally full or empty. This spinless conductivity mechanism is consistent with the absence of any significant Pauli susceptibility in the metallic regime of SbF_5-doped PPP. Furthermore, such a band structure is in agreement with electron-energy-loss spectra for AsF_5-doped PPP. These spectra indicate the appearance

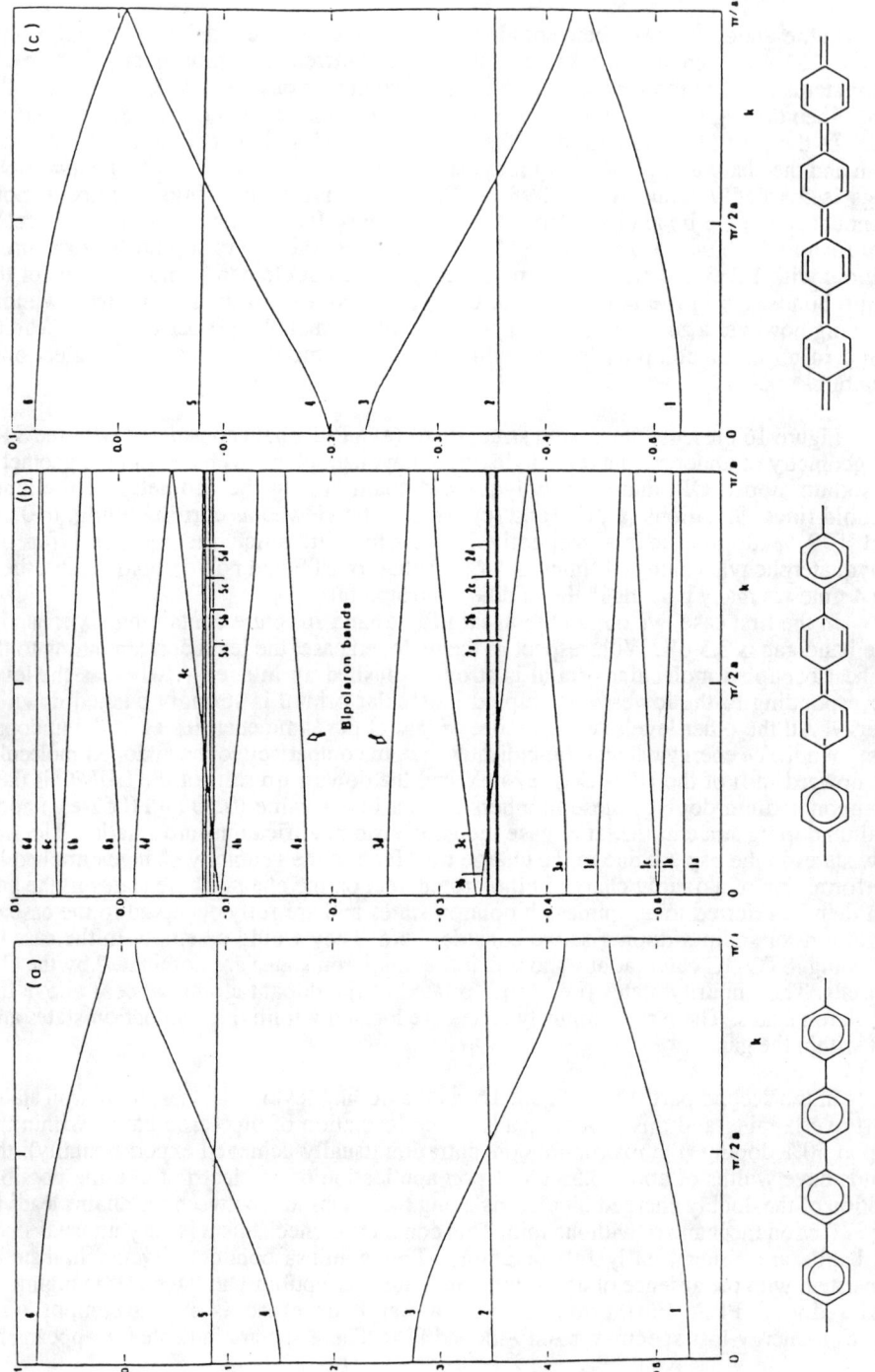

Figure 16. π-Band structures of polyparaphenylene (doped and undoped)

transitions from the valence band to two relatively narrow bands in the gap.

In the third case (c) of Figure 16, we consider a hypothetical doping level of 100% (one dopant per monomer) and a merging of the lower bipolaron band with the valence band and of the higher bipolaron band with the conduction band is obtained. Traditional conductivity with spin could then occur upon n- or p-type doping. There remains a gap of about 1 eV between the original valence band and conduction band.

It can be observed also that the quinoid structure has a lower ionization potential and a larger electron affinity than the benzenoid structure and as a result has a smaller gap. This explains that upon doping the presence of an extra charge on the chain induces a local geometry relaxation from the benzenoid structure towards the quinoid structure. Thus the formation of a charged defect such as a bipolaron actually occur when the lowering in ionization energy due to the presence of a quinoid segment more than compensates for the increase in elastic energy required to form that quinoid segment. Charge transfer processes result in significant local modifications of the chain geometry. In turn, that local geometry modification markedly affects the electronic structure by inducing localized electronic states in the gap. These levels are all attributable to charge-transfer induced modifications of the π system of the polymer and in that sense are intrinsic to the parent material which is the organic polymer. Note that this is in contrast to the situation in doped inorganic semiconductors where the states in the gap are dopant levels. The fact that, upon ionization or electronic excitation, a local relaxation of the lattice geometric and electronic structures is energetically favorable, constitutes the basis of the fascinating physics occuring in doped organic polymers.

5. A THIRD EXAMPLE: MOLECULAR DESIGN OF NON-LINEAR OPTICALLY ACTIVE POLYMERS

The organic solid state has recently gained much interest in the fields of highly conducting polymers and non-linear optics. In the latter, their interest over inorganics is the occurrence of much greater effects due to higher optical damage thresholds, purely electronic effects inducing quasi-instantaneous response and ultra fast signal processing. Their excellent mechanical and molding properties added to the virtually unlimited potentialities of organic synthesis have suscitated many studies and the development of a new physics. They raise interesting questions that will be exemplified by a model study of the linear electric polarizability of a linear chain of hydrogen atoms in the presence of an external field and generalized to polyacetylene chains.

In the study of the perturbation due to the switching of an external electric field, it is anticipated that the polarizability, normalized to the monomeric units, tends to reach an asymptotic limit which should grow when the systems exhibit increased geometrical regularity (metallic situation). For complex systems, this limit will soon be out of reach from studies on chains of increasing length. Thus, it would be very useful to be able to estimate this limit from calculations on infinite chains. One could consider as being rather trivial replacing field-dependent MO's by field-dependent Bloch polymeric orbitals and assuming the usual periodicity properties. However, two types of questions are raised.

On the one hand, as shown by Churchill and Holmstrom [43,44], serious difficulties arise in imposing realistic boundary conditions to solve the one-electron eigenvalue equations; under the boundary conditions commonly used in treating the zero-field case (e.g., Born- Karman boundary conditions), this equation either leads to physically inconsistent results or, still worse, has no solution at all. This strange behavior is a consequence of the pathological nature of the perturbing term, e**F**.**r**, due to the external electric field **F** which becomes undetermined in the limits as **F**→0 and **r**→∞.

On the other hand, the periodic character of the perturbation is not guaranteed under the non-periodic linear external perturbation which would rule out the use of field-perturbed Bloch orbitals. That point has been investigated by Finite-Field calculations over a chain of 24 hydrogen atoms. The results are summarized in Figure 17 which presents the net charge in the unperturbed and perturbed systems. It is seen that the response appears at first sight to be periodic (at least in the middle of the molecule) even if the resulting potential deviates from the ideal e**F**.**r** behavior. An ab initio study of the asymptotic limit of the infinite one-dimensional periodic chain is being developed in our laboratory by Barbier [45] using the SOS (Summation-Over-States) perturbative methodology of Genkin & Mednis [46].

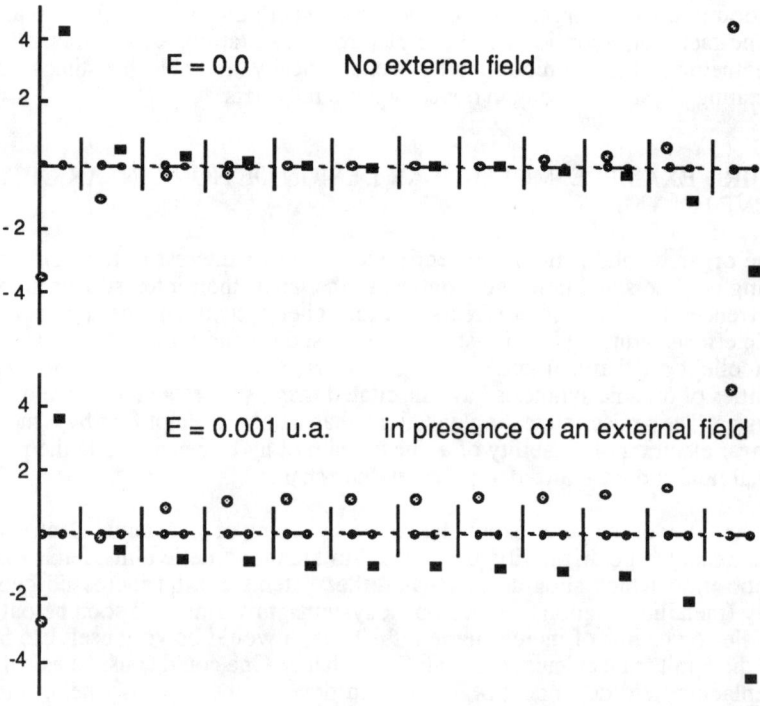

Figure 17. Net charges (x 10^{-2}) of a linear chain of 24 hydrogen atoms in absence (top) and in the presence (bottom) of an external electrical field obtained by 4-31G calculations

The size-dependency of the electronic polarizabilities can be illustrated in the polyene series. For regular polyene chains, both free-electron and Hückel theories predict a behaviour of the longitudinal polarizability proportional to L^3 (where L is the total chain length). Accordingly, the longitudinal polarizability per unit cell grows as the second power of the chain length, L^2, and diverges in the limit of an infinite chain. However, both free-electron and simple Hückel models do not take into account Coulombic interactions explicitly. An investigation of the effect of size and bond alternation on oligomers of various types has been explored to get an indication on the saturation pattern of (hyper)polarizabilities by a more elaborate methodology which includes electron-electron interactions in an average way and allows for the electron charge distribution to relax self-consistently as the size of the system increases.

The Finite-Field (FF) STO-3G results [47-49] are schematized in Figure 18 for several bond alternation degrees and sizes of the main chain. The investigation of the dependence of the longitudinal (FF) polarizability on bond- and chain-length modifications shows an increase of the polarizability by unit cell with respect to size and a decrease with respect to larger bond alternation. Similar trends are observed in the SOS calculation of the longitudinal cubic hyperpolarizability.

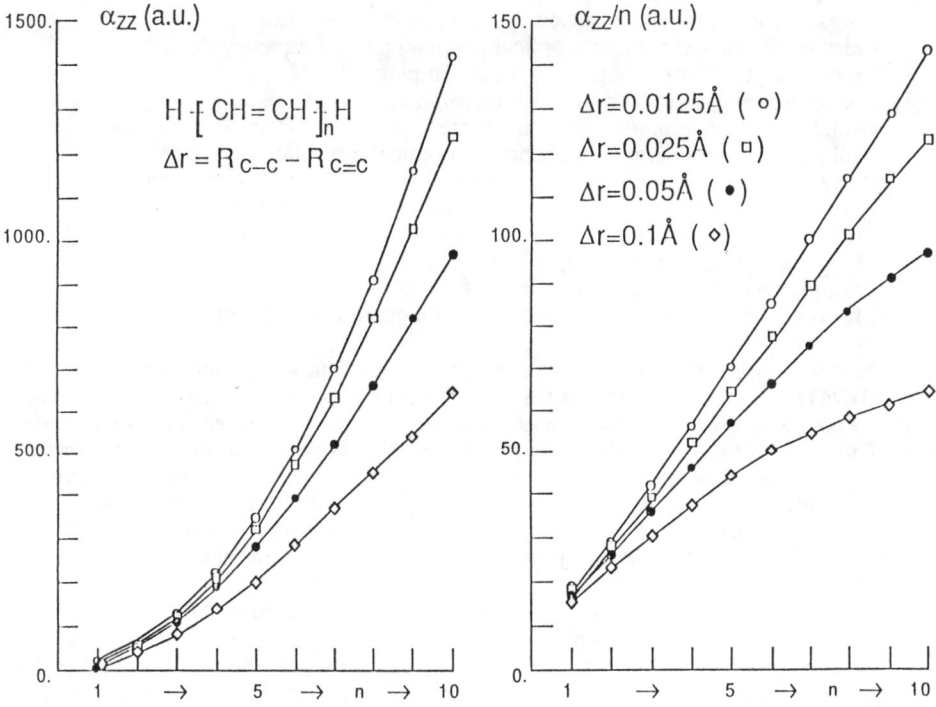

Figure 18. Longitudinal polarizabilities α_{zz} (total [left] and per CH=CH unit [right] of polyenes as a function of the number (n) of CH=CH units

A detailed analysis of the results shows that structural alternations, enforced chemically or mechanically, and ranging from 0.05 to 0.12 Å are predicted to yield important changes in α_{zz} and suggests the possibility of enhancing the value of not only the polarizability α but also the hyperpolarizabilities β and γ by controlled tuning of the geometry. It also raises some warnings in the use of bond additivity schemes when applied to (hyper)polarizabilities considering their high sensitivity upon relatively small bond length differences. Theoretical studies have been published on polydiacetylenes [47], polyallenes and cumulenes[50,51], conjugated chains isoelectronic of octatetraene [52], hydrogen-bonded dimers [53].

6. CONCLUSIONS.

The previous examples should have convinced the reader that there exists now a new intermediate field between physics and chemistry which could be called polymer chemical physics, which is open to calculation and where computational polymer quantum chemistry has to play a great role in the near future.

In a book devoted to "strategies for computer chemistry", it is worth to remind the numerous fields of electronics to which the organic solid state and its theoretical analysis described in parts 4 & 5 of this paper could contribute:
- organic metals ↔ electronic components, plastic batteries,
- electro-optics and nonlinear optical phenomena ↔ frequency doublers, modulators, integrated optics, optical computers,

or other timely fields for an active research as:
- solid-state reactions ↔ chemical sensors,
- solid-state photochemical reactions ↔ optical information storage,
- piezoelectric or ferroelectric phenomena ↔ transducers (electret microphones),
- organic photoconductors or semiconductors ↔ photocopiers, solar cells,
- organic superconductors ↔ Josephson junctions, computer logic gates, high-field magnets, generator, motors,
- liquid crystals ↔ electronic displays,
- ferromagnetism ↔ magnetic recording, magneto-optic recording.

As conclusions, it is also interesting to report the famous quotation by A.Quételet (1796-1874) in "Instructions populaires sur le calcul des probabilités, Tarlier, Bruxelles, 1828" : *Plus les sciences se perfectionnent, plus elles tendent à rentrer dans son domaine (du calcul), qui est une espèce de centre vers lequel elles viennent converger. On pourrait même juger du degré de perfection auquel une science est parvenue par la facilité plus ou moins grande avec laquelle elle se laisse aborder par le calcul, ce qui s'accorde avec ce mot ancien qui se confirme de jour en jour: mundum numeri regunt.* translated in English in ref.[54]:" The more progress physical sciences make, the more they tend to enter the domain of mathematics, which is a kind of centre to which they all converge. We may even judge the degree of perfection to which a science has arrived by the facility to which it may be submitted to calculation". It is complemented by P.A.M.Dirac (1902-1984) : "The underlying physical laws necessary for the mathematical theory of a large part of physics and the whole of chemistry are thus completely known, and the difficulty is only that the exact application of these laws leads to equations much too complicated to be soluble".

Perhaps, the truth is due to A.Comte (1798-1857) in his "cours de philosophie positive, tome troisième, contenant la philosophie chimique et la philosophie biologique,

Bachelier, Paris, 1838, pp.41-42" : *Toute tentative de faire rentrer les questions chimiques dans le domaine des doctrines mathématiques, doit être réputée jusqu'ici, et sans doute à jamais, profondément irrationnelle, comme étant antipathique à la nature des phénomènes......si, par une aberration heureusement presque impossible, l'emploi de l'analyse mathématique acquérait jamais, en chimie, une semblable prépondérance, il déterminerait inévitablement, et sans aucune compensation, dans l'économie entière de cette science, une immense et rapide rétrogradation......* or, in the previously cited translation[54]: "Every attempt to employ mathematical methods in the study of chemical questions must be considered profoundly irrational and contrary to the spirit of chemistry.... If mathematical analysis should ever hold a prominent place in chemistry -an aberration which is happily almost impossible- it would occasion a rapid and widespread degeneration of that science.

ACKNOWLEDGEMENTS

This paper is a contribution presented at the seminar on "Strategies for Computer Chemistry" held in Novara (Italy). The author is very grateful to Prof. Camillo TOSI for allowing him to attend that stimulating experience. He is also indebted to Prof. Joseph DELHALLE and Drs. Joseph FRIPIAT, Jean-Luc BREDAS, and Christian BARBIER for frequent and helpful discussions on the topics developed in this paper. All calculations reported here have been made on the Namur-Scientific Computing Facility (Namur-SCF), a result of a cooperation between the Belgian National Foundation for Scientific Research (FNRS), IBM-Belgium and the Facultés Universitaires Notre-Dame de la Paix (FUNDP).

REFERENCES

[1] R.Olby, J.Chem.Educ, **47**, 168 (1970)
[2] F.W.Billmeyer, in **Textbook of Polymer Science**, 2nd edition, J.Wiley (1971)
[3] Chemical & Engineering News, August 24, 1987, p.27
[4] National Research Council, **Opportunities in Chemistry**, Pimentel Report, National Academy Press, Washington DC (1985), p.49
[5] Chemical & Engineering News, June 8, 1987
[6] W.L.McCubbin, R.Manne, Chem.Phys.Letters, **2**, 230 (1968)
[7] J.M.André, G.Leroy, Chem.Phys.Letters, **5**, 71 (1971)
[8] B.Pullman, A.Pullman, **Quantum Biochemistry**, Interscience Publishers (1963), see pp.304 & 305
[9] J.M.André, **Etude Théorique de la Structure de Bandes des Systèmes Périodiques**, Ph.D.Thesis, University of Louvain (UCL), Belgium, (1968) see pp.131 & sq.
[10] J.M.André, in **Electronic Structure of Polymers and Molecular Crystals**, J.M.André, J.Ladik (Eds.), NATO-ASI Series B9, 1, Plenum Press (1975)
[11] J.M.André, J.Delhalle, in **Quantum Theory of Polymers**, J.M.André, J.Delhalle, J.Ladik (Eds.), NATO-ASI Series C39, 1, D.Reidel Publishing Company (1978)
[12] J.M.André, Advan.Quantum Chem. **12**, 65 (1980)
[13] J.M.André, in **Current Aspects of Quantum Chemistry**, R.Carbo (Ed.), 273, Elsevier (1982)
[14] R.Hoffmann, R.B.Woodward, J.Amer.Chem.Soc. **87**, 2046 (1965)

[15] J.P.Lowe, S.A.Kafafi, J.P.LaFemina, J.Phys.Chem., **90**, 6602 (1986)
[16] K.Tanaka, S.Yamashita, H.Yamabe, T.Yamabe, Synth.Metals, **17**, 143 (1987)
[17] J.M.André, L.Gouverneur, G.Leroy, Internat.J.Quantum Chem., **1**, 427 (1967)
[18] J.M.André, J.Chem.Phys.,**50**, 1536 (1969)
[19] J.M.André, V.P.Bodart, J.L.Brédas, J.Delhalle, J.G.Fripiat, in **Quantum Chemistry of Polymers - Solid State Aspects**, J.Ladik, J.M.André & M.Seel (Eds.), NATO-ASI Series, C**123**, 1, D.Reidel Publishing Company (1984)
[20] J.M.André, D.P.Vercauteren, G.B.Street, J.L.Brédas, J.Chem.Phys.,**80**, 5643 (1984)
[21] J.Delhalle, L.Piela, J.L.Brédas, J.M.André, Phys.Rev., B**22**, 6254 (1980)
[22] J.M.André, D.P.Vercauteren, V.P.Bodart, J.G.Fripiat, J.Comput.Chem., **5**, 535 (1984)
[23] J.M.André, L.Gouverneur, G.Leroy, Bull.Soc.Chim.Belges, **76**, 661 (1967)
[24] J.M.André, J.Delhalle, C.Demanet, M.E.Gérard-Lambert, Internat.J.Quantum Chem., S**10**, 99 (1976)
[25] C.R.Ginnard, W.M.Riggs, Anal.Chem., **44**, 1310 (1972)
[26] M.H.Wood, M.Barber, I.H.Hillier, J.M.Thomas, J.Chem.Phys., **56**, 1788 (1972)
[27] J.M.André, J.Delhalle, Chem.Phys.Letters., **17**, 145 (1972)
[28] J.Delhalle, D.Thelen, J.M.André, Comput.Chem., **3**, 1 (1979)
[29] J.L.Brédas, J.M.André, J.Delhalle, Chem.Phys., **45**, 109 (1980)
[30] J.M.André, J.Delhalle, J.J.Pireaux, in **Photon, Electron, and ion Probes of Polymer Structure and Properties**, D.W.Dwight, T.J.Fabish, H.R.Thomas, Eds., ACS Symposium Series, **162**, 151 (1981)
[31] R.G.Snyder, J.Chem.Phys., **47**, 1316 (1967)

[32] J.J.Pireaux, J.Riga, R.Caudano, J.J.Verbist, J.Delhalle, S.Delhalle, J.M.André, Y.Gobillon., Phys.Scripta, **16**, 329 (1977)
[33] H.Kuhn, Chimia AArau, **2**, 1 (1948)
[34] H.Kuhn, Helv.Chim.Acta, **31**, 1441 (1948)
[35] H.Kuhn, J.Chem.Phys., **17**, 1198 (1949)
[36] J.Delhalle, J.L.Brédas, J.M.André, Chem.Phys.Letters, **78**, 93 (1981)
[37] J.L.Brédas, B.Thémans, J.M.André, R.R.Chance, D.S.Boudreaux, R.Silbey, Journal de Physique, **44**, C3, 273 (1983)
[38] J.A.Pople, S.H.Walmsley, Molec.Phys. **5**, 15 (1962)
[39] W.P.Su, J.R.Schrieffer, A.J.Heeger, Phys.Rev., B**22**, 2099 (1980)
[40] J.L.Brédas, R.R.Chance, R.Silbey, Phys.Rev., B**26**, 5843 (1982)
[41] J.L.Brédas, B.Thémans, J.M.André, Phys.Rev., B**26**, 6000 (1982)
[42] J.L.Brédas, B.Thémans, J.G.Fripiat, J.M.André, R.R.Chance, Phys.Rev., B**29**, 6761 (1984)
[43] J.N.Churchill, F.E. Holmstrom, Amer.J.Phys., **50**, 848 (1982)
[44] J.N.Churchill, F.E. Holmstrom, Physica B.,**123**, 1 (1983)
[45] C.Barbier, to be published
[46] V.M.Genkin, P.M.Mednis, Sov.Phys.-JETP **27**, 609 (1968)
[47] V.P.Bodart, J.Delhalle, J.M.André, Springer Series in Solid State Sciences, **63**, 191 (1985)
[48] J.M.André, C.Barbier, V.P.Bodart, J.Delhalle, in **Nonlinear Optical Properties of Organic Molecules and Crystals**, D.S.Chemla & J.Zyss (Eds), Vol.2, 137, Academic Press (1987)
[49] J.M.André, J.O.Morley, J.Zyss, in **Volume in Tribute of Prof. R.Daudel**, in press (Reidel Publishing Company)

[50] J.Delhalle, V.P.Bodart, M.Dory, J.M.André, J.Zyss, Internat.J.Quantum Chem., S**19**, 313 (1985)
[51] V.P.Bodart, J.Delhalle, M.Dory, J.G.Fripiat, J.M.André, J.Optical Soc.Amer., B**4**, 1047 (1987)
[52] E.Younang, J.Delhalle, J.M.André, Nouveau J.Chimie, **19**, 403 (1986)
[53] M.Dory, J.Delhalle, J.G.Fripiat, J.M.André, Internat.J.Quantum Chem., accepted for publication
[54] W.J.Hehre, L.Radom, V.R.Schleyer, J.A.Pople, in **Ab initio Molecular Orbital Theory**, J.Wiley, New-York (1985)

Ab Initio Configuration Interaction Study of Electronic and Geometric Structure of Alkali Metal Clusters

Piercarlo Fantucci
Dipartimento di Chimica Inorganica e Metallorganica
Universita di Milano, Via Venezian 21, 20133 Milano, Italy

Vlasta Bonačić-Koutecký and Jaroslav Koutecký
Institut für Physikalische und Theoretische Chemie
Freie Universität Berlin, Takustr. 3, 1000 Berlin 33, FRG

I. INTRODUCTION

The investigation of electronic and geometric properties of metal clusters employing quantum chemical methods is a challenging task from a theoretical point of view and represents an interesting field of applicability of the modern computational chemistry techniques.

In order to gain the understanding about general rules governing the nature of the chemical bond occuring in clusters, the chosen theoretical method should account for the very specific electronic properties of metal clusters.

In the literature, several different proposals have been made concerning the theoretical determination of the stability of clusters. One of them considers that the position of the nuclei is the dominant factor determining the stability: Therefore, the two-body (or many-body) potentials can be conveniently used to determine the cluster stability |1,2|. The second method broadly used in cluster science is based on electronic shell concept: It neglects completely or partly, the importance of the position of the nuclei (jellium model) |3|. In the customary quantum chemical approaches the importance of the framework of nuclei and the electron density, as well as their mutual dependence, can be considered explicitely. Obviously, the large number of electrons present in heavier alkali metals represents a limitation to the applicability of this approach. Therefore, it is natural to carry out first a systematic investigation of Li_n clusters, employing ab initio configuration interaction (CI) method (Ref. 4 and references therein). Several simple general rules governing the geometric and electronic structure of small Li clusters can be formulated. They can be summarized as following |5|:

1. Compactness of the cluster favours the stability, giving rise usually to the highly symmetrical form.
2. The Jahn-Teller or pseudo-Jahn Teller effect is responsible for deformation of the cluster geometries towards less symmetrical shapes. For instance, the planar forms of small clusters have smaller number of degenerated molecular orbitals (MO) or natural orbitals (NO) than the highly symmetrical three dimensional structures |4-6|.
3. The topological properties can play an important role, since the chemical bond in the lithium clusters is relatively delocalized |7|.
4. The role played by the 2p functions promoting the s-p hybridization of the lithium atoms is of fundamental importance in determining the geometrical and electronic properties of small Li_n clusters |8-10|.

Even very approximate theoretical methods can give reliable predictions about lithium cluster properties provided they are able to account for these underlying principles.

The question can be raised if these rules remain valid for other alkali metal clusters. The next step is the consideration of Na_n clusters which have been already investigated with the local spin density (LSD) method |11-13|. It is important to note that the optimal shapes of the Na_n clusters obtained by the LSD study were in qualitative agreement with the optimal forms of the Li_n clusters obtained from the ab-initio CI work |4|.

Comparison of the electronic and geometric properties of Li_n and Na_n clusters will be presented in this paper. Some important methodological and computational problems inherent to the study of the electronic and geometric properties of the alkali metal clusters will be also considered. The results of parallel investigations carried out at all-electron (AE) and valence electrons only (ECP) in the framework of the Hartree-Fock (HF) procedure for optimum geometry search of Na_n clusters |14| will be discussed. As well known, the CI methods making use of a truncated expansion suffers from the "size inconsistency" error. The computed correlation energy is not an additive properties with the number of electrons. Since the inclusion of correlation effects is crucial for the determination of the stability of clusters, the MRD-CI or direct CI calculations have been carried out for the HF optimum cluster shapes. The size consistency error has been estimated according to the Linear Coupled Pairs Method (LCPMET) method |15,16| in order to examine the reliability of the cluster stabilities obtained from the above mentioned CI treatments.

II. Computational Methods.

The geometries of neutral Li_n and Na_n clusters have been optimized using an analytical gradient method in the framework of all electron Hartree-Fock procedure (AE-HF) |17|. In the case of Na_n the geometry-energy search has been carried out also with effective-core-potential Hartree-Fock approximation (ECP-HF) employing a numerical algorithm |18|. For the optimized AE-HF and ECP-HF geometries the energies have been determined taking into account correlation effects employing multireference double-excitation configuration interaction procedure (MRD-CI) |19,20|. Only valence electrons have been correlated. The direct CI procedure |21,22| has been employed for several important cases in order to check the reliability of the extrapolation technique used in the MRD-CI procedure. Davidson correction for the full CI energy has been calculated. In order to estimate the size consistency error one reference configuration coupled-cluster method has been applied |15,16|.

Relatively small AO basis sets have been used taking into account appropriate polarization functions. In the case of Li_n the minimal basis set augmented by one p-function has been used since it was shown that extension of the basis sets does not substantially change the results |4|. In the case of Na_n, the (9s,4p) AO basis set contracted to (3s,1p) |23,24| and augmented by one diffuse p function with exponent $\alpha_p = 0.065$ has been used for the AE calculations. Careful investigations of the tetramer properties with different larger AO basis sets has confirmed again the reliability of the chosen relatively small basis set |14|. For HF-ECP calculations of Na_n the definition of the effective core potential and the basis set for valence 3s orbital are adopted from Ref. 25. The basis is augmented by the same diffuse p function as in the case of the AE treatment.

III. The Geometries of Li and Na Clusters

The shapes of small Li and Na clusters developing with the increasing nuclearity are parallel in details (Figs. 1 and 2). Li_4, Li_5, Na_4 and Na_5 are slightly distorted segments of the (111) plane of the fcc crystal lattice. The ground state energies of the pentagonal pyramidal (C_{5v}) forms of Li_6 and Na_6 are degenerated with the ground state energies of the respective planar Li_6 and Na_6 (D_{3h}) shapes which can be considered as segments of the fcc crystal lattices. Li_7 and Na_7 might be taken as a step in the pentagonal cluster growth. Li_8 and Na_8 have highly symmetrical T_d shapes with four pyramids built on the four faces of an inner tetrahedron. The ninth Li atom in Li_9 can be added perpendicular to one side of the inner tetrahedron of the Li_8 T_d. This Li_9 structure is topologically equivalent to the pentagonal bipyramid with two additional atoms capping two neighbouring faces of one of its pyramidal subunits. The energy of this Li_9 isomer is almost equal to the energy for the Li_9 structure built by capping two neighbouring faces of different pyramidal substructures. Several Li_9 isomers are energetically close lying. The topology of the optimum geometry for Na_9 corresponds to the topology of the second stable Li_9 isomer. The only larger difference between Li and Na geometries occurs for heptamers. Although

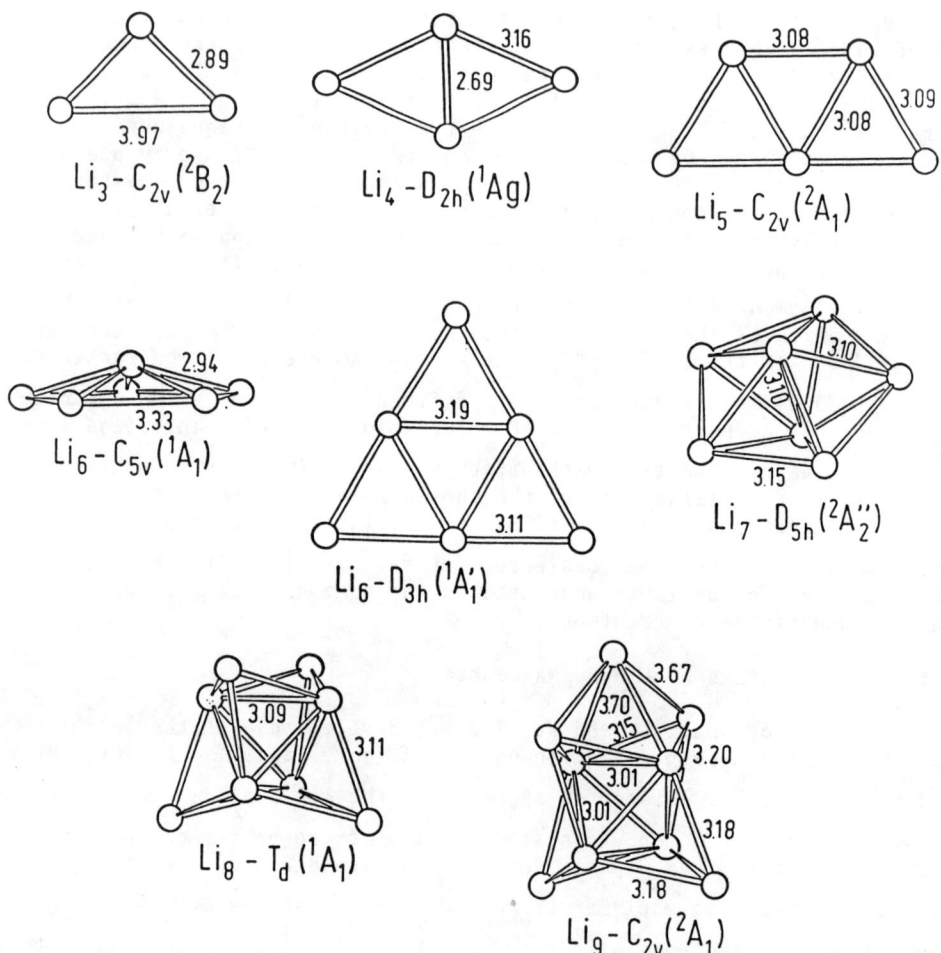

Fig. 1: Geometries of energy optimized Li clusters. The interatomic distances (Å) resulting from the SCF optimal geometry search followed by the partial MRD CI optimization ("scaling" procedure described) are shown. The symmetry group of the cluster as well as the irreducible representation and the spin multiplicity of the respective state (in the bracket) are given.

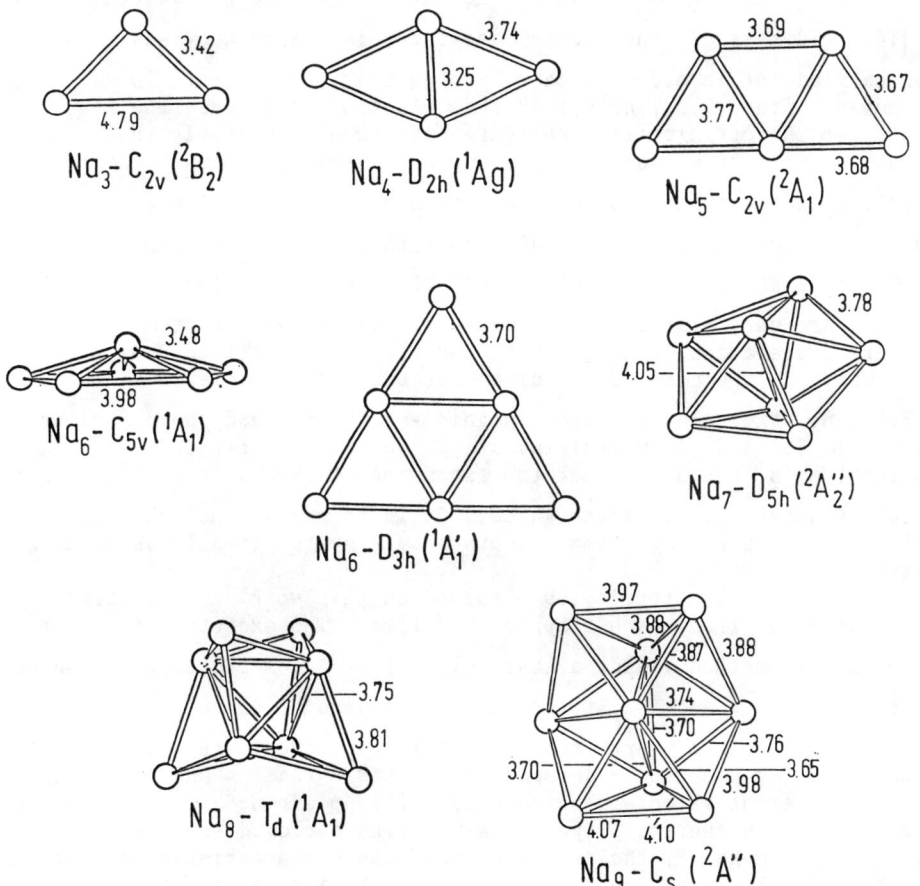

Fig. 2: Topologies and geometries of optimized Na clusters. Bond lengths obtained at AE-HF level are given. Very similar values have been obtained with ECP-HF and ECP-CI geometry optimization.

the pentagonal bipyramid (D_{5h}) appears to be most stable structure for Li_7 and Na_7, the distance between the apical atoms in the D_{5h} Na_7 cluster is larger than the sides of its pentagon which is not the case for Li_7. Therefore Li_7 pentagonal bipyramid consists of five condensed tetrahedra. Nevertheless, in the case of Na_7 the minimum is very shallow i.e. the change in the distance between the apical atoms does not influence the energy substantially.

It is necessary to emphasize that the optimum geometry-energy search has been carried out much more extensively for Li_n |4| than for Na_n |14| clusters. In the former case the normal frequency analysis has been carried out in order to check if the stationary point is a real minimum on the global energy surface. In the latter case the starting points for the optimization procedure have been used mostly in analogy to the Li_n clusters. Nevertheless, we believe that no important stable structures for the Na_n clusters have been missed. The distances in Na_n clusters are of course in general larger than in Li_n clusters.

The AE-HF, ECP-HF as well as ECP-CI interatomic distances for Na_n clusters are very similar. The interatomic distances reported in this work are systematically larger than the distances obtained from the LSD procedure. In the case of Na_2 dimer the LSD approach yields a distance of 5.5 a.u., the value computed in this work is 6.1 and the experimental one is 5.8 a.u. A careful investigation of the interatomic distances for the Na_4 has shown that the extension of the basis set does not influence substantially these values. It is to expect that the consideration of core polarization effects should systematically shorten the distances.

In general the topologies obtained in this work are similar to those obtained from the LSD approach |11-13|. The exception is Na_8 for which the LSD method yields a less symmetrical D_{2d} structure. The high symmetrical Li_8 and Na_8 T_d structures obtained from the ab initio CI calculation have two very special characteristics: the tetrahedral subunits do not need to be deformed from steric reasons and simultaneously the eight valence electrons can fill up a non-degenerate s-type MO and triply degenerate p-type MO's yielding a closed shell electronic structure. In general, the development of the characteristic geometries with the size for the most stable neutral alkali metal clusters exhibit the following regularities: Very small clusters (n=3,...6) are deformed sections of the (111) plane of the fcc crystal lattice. The geometries of the most stable heptamer, octamer and larger clusters are condensed deformed tetrahedra appropriately sharing their triangular faces. Starting from heptamers the fcc topologies are less favourable than the clusters built from tetrahedra.

IV. The stability of Li and Na clusters.

The atomization energies per atom

$$E_b/n = E_1 - E_n/n$$

as a function of the cluster nuclearity is drawn in Fig. 3 for Li and Na clusters. The comparison of the curves obtained from different

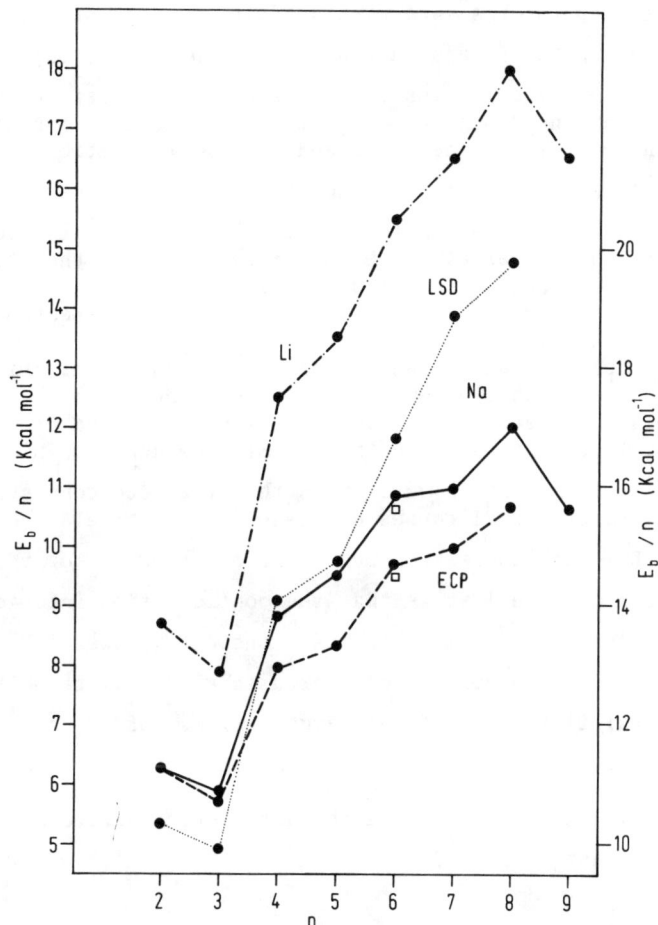

Fig. 3: The atomization energy E_b/n as a function of the number of atoms in the Li and Na clusters. The broken lines (-.-) labels the all-electron MRD-CI results for Li clusters. The solid (___) and dashed (---) lines label the all-electron MRD-CI and effective potential MRD-CI results for Na clusters, respectively. The energies are determined for the corresponding HF optimal geometries. The dotted curve labels LSD results (Ref. 11-13) for Na clusters for which the scale on the right-hand side is valid. The ECP-MRD-CI energies obtained for the ECP-CI optimized geometries differ so little from the ECP-MRD-CI energies for the ECP-HF optimized geometries that they cannot be distinguished from the latter on the scale in this figure.

theoretical approaches are displayed in the case of the Na_n clusters. The E_b/n curves for Li and Na clusters exhibit striking similarities. The values of atomization energy generally increase with the cluster size. The pronounced minima for trimers and maxima for octamers are present in both cases. The AE-CI and ECP-CI atomization E_b/n curves for Na_n are completely parallel. The LSD atomization energy curve |11-13| is steeper with increasing cluster size and in general the values of the binding energy per atom are considerably larger then those obtained in this work. However, the use of local approximation in density functional theories can give rise to a rather overestimated cohesive energy.

In order to examine the reliability of the atomization energy as a function of the cluster nuclearity obtained from the CI treatment, the correlation energies per atom (electron) obtained from the SD-CI, Davidson full CI estimate and the LCPMET procedure for Li_n clusters are given in Table 1. It is well known that the size consistency error in the single reference CI causes decrease of the correlation energy per electron E_c with increasing number of electrons n_e of the order $n_e^{-1/2}$ for larger n_e. Table 1 shows that the configuration interaction yields for Li_n cluster n=2-9 more or less constant values of E_c. The E_c calculated with Davison correction increases with n_e. The LCPMET procedure gives similar increase but larger values of E_c's. Nevertheless, the LCPMET method represents the upper limit of the correlation energy. Therefore it is possible to conclude that the increase of the correlation energy per electron is a characteristic feature of the small Li_n cluster at least up to $n \leq 9$.

Table 1: Correlation energy per electron for Li_n clusters in kcal/mole.

Li_n	E_c (ext)	E_c (Dav.)	E_c (LCPMET)
2	7.25	7.31	7.88
3	6.83	7.5	7.7
4	7.99	9.11	9.66
5	7.20	8.36	10.69
6 (D_{3h})	7.68	8.97	9.81
6 (C_{5v})	8.2	9.61	10.51
7	8.4	9.88	10.88
8	7.81	9.35	10.97
9	8.38	10.06	11.47

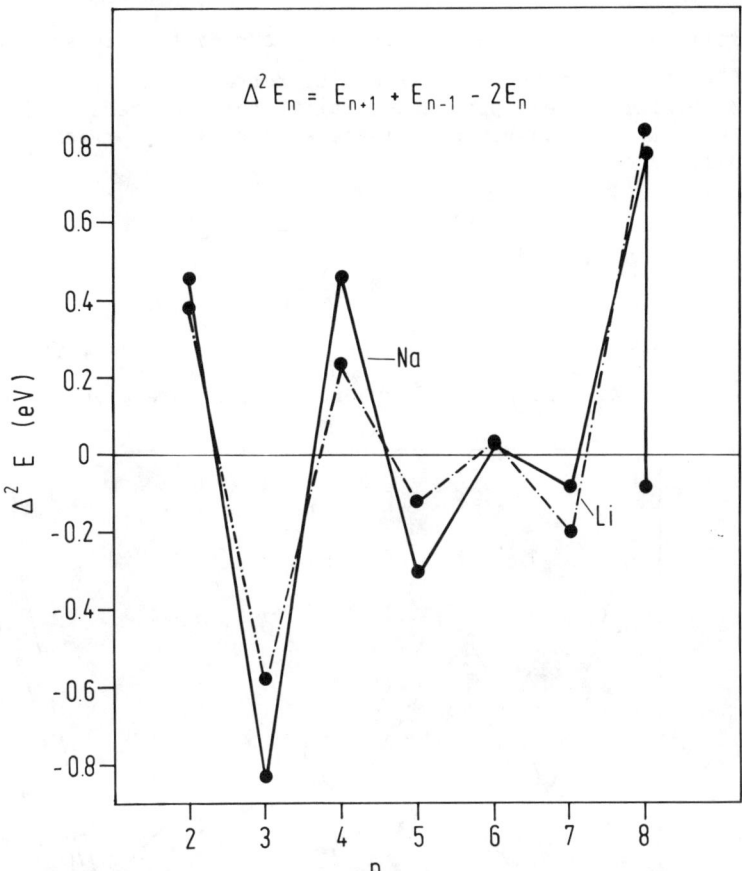

Fig. 4: The second differences $\Delta^2 E_n$ as a function of the number of Li (-.-) and Na (___) clusters. The results are derived from the AE MRD-CI energies calculated for the AE-HF optimized geometries

The second difference of the cluster energy $\Delta^2 E_n$ as a function of the cluster size for Li_n and Na_n is shown in Fig. 4. The quantity $\Delta^2 E_n$ defined as

$$\Delta^2 E_n = E_{n+1} - E_{n-1} - 2E_n = (E_{n-1} + E_1 - E_n) - (E_n + E_1 - E_{n+1})$$

can be interpreted as the difference between the fragmentation energies of X_n and X_{n+1} to give the X_{n-1} and X_n clusters, respectively. Again

the striking similarities between $\Delta^2 E_n$ curves for Li_n and Na_n clusters can be found. Pronounced maxima for tetramers and octamers indicating their stability, less pronounced maxima for hexamers and minima for clusters with odd number of valence electrons are present.
The differences in energy

$$\delta E_n = E_n - (E_{n-1} + E_1)$$

$$\delta' E_n = E_n - (E_{n-2} + E_2)$$

characterizing the dissociation channel I ($X_n \rightarrow X_{n-1} + X$) and II ($X_n \rightarrow X_{n-2} + X_2$), respectively where X=Li, Na are given in Fig. 5.

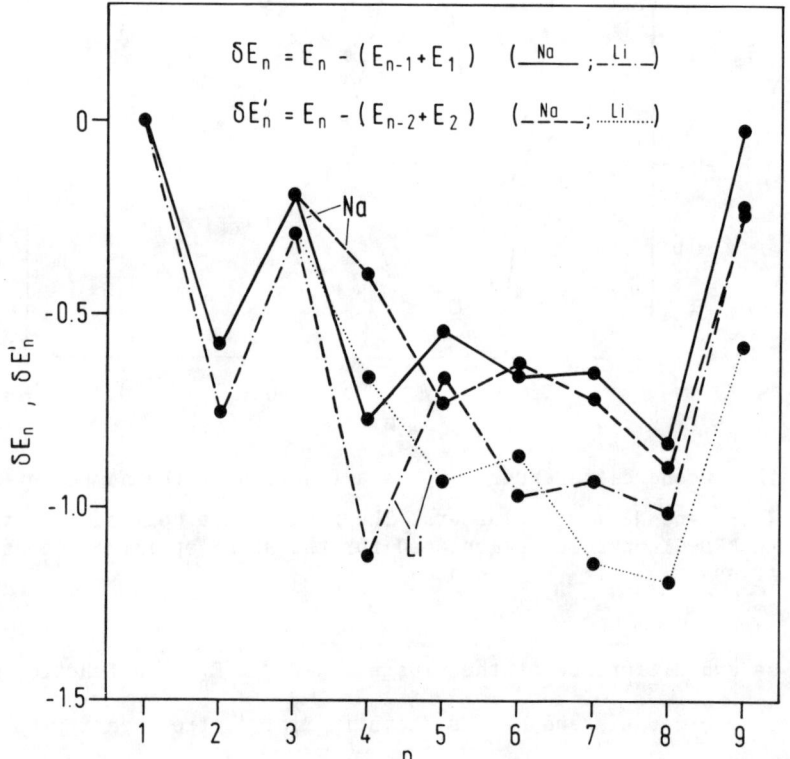

Fig. 5: Dissociation energies δE_n and $\delta' E_n$ for channel I and II: $Li_n \rightarrow Li_{n-1} + Li$ (-.-); $Na_n \rightarrow Na_{n-1} + Na$ (___), $Li_n \rightarrow Li_{n-2} + Li_2$ (...); $Na_n \rightarrow Na_{n-2} + Na_2$ (---). The results are derived from the AE MRD-CI energies calculated from the AE-HF optimized geometries.

The maxima of δE_n for trimers and nonamers show that the dissociation in X_2+X_1 and X_8+X_1 is favourable. Comparing δE_n with $\delta' E_n$ one can conclude that dissociation of tetramers into dimers is more favourable than into trimers since the trimers are characterized by a low stability. For n=5, 7, 8 and 9 the channel I is more favourable than the channel II but for n=6 the dissociation into tetramer and dimer is again slightly more favourable.

The calculated vertical ionization potentials for Li_n and Na_n are shown in Figure 6. A completely parallel behaviour is striking. The characteristic features are even-odd oscillations and the maxima for n=4 and 8. The values of ionization potentials for Na_n are systematically lower than the experimental values |26,28| by about of 0.3 - 0.5 eV although the general trend is parallel. After careful investigation of the influence of the AO basis sets as well as of the correlation effects for valence electrons only, the conclusion has been reached that the core polarization effects might be important for determination of ionization potentials and should be taken into account.

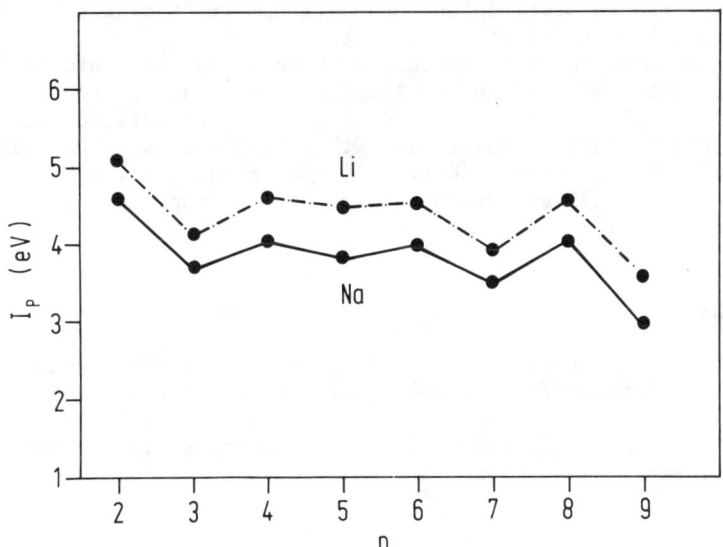

Fig. 6: Vertical ionization potentials (IP's) in (eV) for Li_n (-.-) and for Na_n (___) clusters obtained from the AE-MRD-CI calculations.

V. CONCLUSIONS

The results obtained from Hartree-Fock and configuration interaction studies in both all-electron and effective-core-potential versions, show that the most important electronic properties of sodium clusters are in full accordance with the corresponding properties of the lithium clusters. The similarity between the valence-electron shell of Li and Na atoms plays here the most important role. Although the theoretical treatment of Na clusters is much more difficult than for Li clusters and the core-polarization correction should influence properties of Na_n more than Li_n, the results obtained presently offer a consistent picture of the geometrical and electronic structure of small alkali metal clusters as well as of their stability.

The optimized geometries of Li and Na clusters conform with the four rules mentioned in the Introduction (influence of the compactness, of the nodal properties of MO's or NO's, of the topology and of the polarization functions). The geometrical structures with highest symmetry are unfavourable for clusters with unsufficient number of electrons which are available to fill up the degenerated one-electron functions. In this manner it is possible to understand why the planar structures are more stable for very small clusters $n<6$ and three-dimensional shapes built from tetrahedral subunits start to be more stable for $n>7$. The highest stability of alkali metal rhombic tetramers and highly symmetrical "closed shell" T_d octamers suggests the high abundances for these nuclearities in the detection devices if the experimental conditions do not lead to cluster fragmentation |29,30|.

In summary, the quantum chemical methods which take into account the correlation effects yield an explanation of many characteristic properties of small alkali metal clusters from an unified point of view. This approach contains naturally the features of the simple electron shell model but yield additional aspects and results which are inevitable for the full understanding of the electronic properties of the small alkali metal clusters.

Acknowledgement:

This work has been supported by the Deutsche Forschungsgemeinschaft (Sfb 337 "Energy transfer in molecular aggregates") and the Centro Nazionale delle Ricerche (CNR), Rome. The calculations have been carried out on CONVEX C1 and CRAY XMP 24 computers. The authors are grateful to Mr. Boustani for computational help.

References:

|1| M.R. Hoare, Adv. Chem. Phys. 40, 49 (1979).
|2| M.R. Hoare and P. Pal, Adv. Phys. 20, 161 (1971).
|3| W.D. Knight, K. Clemenger, W.A. De Heer, W.A. Saunders, M. Chou and M.L. Cohen, Phys. Rev. Lett. 52, 2141 (1984).
|4| I. Boustani, W. Pewestorf, P. Fantucci, V. Bonačić-Koutecký and J. Koutecký, J. Phys. Rev. B 35, 9437 (1987).
|5| J. Koutecký and P. Fantucci, Chem. Rev. 86, 538 (1986).
|6| J. Koutecký and P. Fantucci, Z. Phys. D 3, 147 (1986).
|7| J. Koutecký, D. Plavšić and D. Döhnert, Croat. Chim. Acta 56, 451 (1983).
|8| P. Fantucci and P. Balzarini, J. Mol. Catal. 4, 337 (1978).

|9| H.-O. Beckmann, J. Koutecký, P. Botschwina and W. Meyer, Chem. Phys. Lett. **67**, 119 (1979).

|10| H.-O. Beckmann, J. Koutecký and V. Bonačić-Koutecký, J. Chem. Phys. **73**, 5182 (1980); D. Plavsic, J. Koutecky, G. Pacchioni and V. Bonačić-Koutecký, J. Phys. Chem. **87**, 1096 (1983); P. Fantucci, J. Koutecký and G. Pacchioni, J. Chem. Phys. **80**, 325 (1984); J. Koutecký and G. Pacchioni, Ber. Bunsenges. Phys. Chem. **88**, 323 (1984); J. Koutecký, G. Pacchioni, G.-H. Jeung and E. C. Hass, Surf. Sci., **156**, 650 (1985).

|11| J.L. Martins, J. Buttet and R. Car, Phys. Rev. B **31**, 1804 (1985).

|12| J.L. Martins, R. Car and J. Buttet, J. Chem. Phys. **78**, 5646 (1983).

|13| J.L. Martins, J. Buttet and R. Car, Ber. Bunsenges. Phys. Chem., **88**, 239 (1984).

|14| V. Bonačić-Koutecký, P. Fantucci anf J. Koutecký, Phys. Rev. B, in press.

|15| J. Cizek, J. Chem. Phys. **45**, 4256 (1966).

|16| unpublished results.

|17| M.F. Guest and J. Kendrick, Daresbury Laboratory Report No. CCP1/86/1, 1986 (unpublished).

|18| M.Y.D. Powell, Comput. J. **7**, 155 (1964).

|19| R. J. Buenker and S. D. Peyerimhoff, Theor. Chim. Acta **35**, 33 (1974); R. J. Buenker, S. D. Peyerimhoff and W. Butscher, Mol. Phys. **35**, 771 (1978).

|20| R.J. Buenker, in Studies in Physical and Theoretical Chemistry: Current Aspects of Quantum Chemistry, edited by R.T. Carbo (Elsevier, Amsterdam, 1982), Vol. 21, pp. 17-34; in Proceedings of the Workshop on Quantum Chemistry Molecule, Physics (Wollongong, Canberra, 1980); R.J. Buenker and R.A. Phillips, J. Mol. Struct. Theochem. **123**, 291 (1985).

|21| P.E. Siegbahn, J. Chem. Phys. **72**, 1647 (1980).

|22| V.R. Saunders and J.H. van Leuthe, Mol. Phys. **48**, 923 (1983).

|23| L. Giaolio, R. Pavani and R. Clementi, Gazz. Chim. Ital. **108**, 181 (1978).

|24| R. Poirier, R. Kari and I. G. Czismadia, Handbook of Gaussian Basis Sets (Elsevier, Amsterdam, 1985).

|25| W.J. Stevens, H. Basch and M. Krauss, J. Chem. Phys. **81**, 6026 (1984).

|26| A. Herrmann, E. Schumacher and L. Wöste, J. Chem. Phys. **68**, 2327 (1978).

|27| K. I. Peterson, P. D. Dao, R. W. Farley and A. W. Castleman, Jr., J. Chem. Phys. **80**, 1780 (1984).

|28| M.M. Kappes, M. Schär, U. Röthlisberger, Ch. Yeretzlav and E. Schumacher, Chem. Phys. Lett. **143**, 251 (1988).

|29| M.M. Kappes, P. Radi, M. Schär and E. Schumacher, Chem. Phys. Lett., **119**, 11 (1985).

|30| C. Bréchignac and Ph. Cahudzac, Z. Phys. **D 3**, 121 (1986).

"WIZARD: Artificial Intelligence and Conformational Analysis"

Daniel P. Dolata
Assistant Professor of Chemistry
UNIVERSITY OF ARIZONA
Department of Chemistry
Tucson, Arizona

The title of this talk is WIZARD, which is also the name of our computer programme. Many people like to give programmes names that mean something. WIZARD doesn't mean anything, I like the name. The WIZARD project has three parts. It combines axiomatic theories, which are mathematical models of reasoning, and artificial intelligence. which is the technique that we utilize to study axiomatic theories. We then apply that to conformational analysis. Those are the three parts of the WIZARD project. Our purpose in the WIZARD project was to provide an expert which the chemists could utilize, sort of a helper to a chemist, so that they could understand the conformational possibilities in a molecule. We wanted a chemist to be able to find <u>all</u> conformers, rather than just one, two, or a large number of them. We also wanted to have the chemist be able to understand such things as the conformational rigidity or flexibility of a molecule.

We wanted to also, at the same time, explore the scope and limitations of artificial intelligence in chemistry. So we are not only looking at a tool to help chemists study conformational analysis, we are also trying to understand the scope and limitation of artificial intelligence in chemistry. And so we constantly have to play this off. We are looking at chemistry, but we are also using chemistry as a test bed for finding the limitations of artificial intelligence. As I am going to show you, a naive approach of using expert systems in chemistry does not fill the bill. Artificial intelligence and expert systems used naively do not give you the results that you need to get. So you have to critically evaluate artificial intelligence and modify it for chemistry, because chemistry is not a simple system. Yet a lot of artificial intelligence was developed in trivial systems such as trying to find a key in a box, looking at little boxes on table tops and so forth. So we have to throw out the advertising, we have to throw out all the garbage that we find in AI, and really put it to work for chemistry. I am going to try and show you how we can do that and how we can make artificial intelligence a good work horse in chemistry.

With that in mind, we started off at the University of Lund with an observation. We observed that a human conformational analysis expert was capable of making good models of compounds. If I draw a totally planar structure of a steroid on the blackboard, then in your mind you can all create a three dimensional diagram of it. This three dimensional diagram is not going to be accurate to a hundredth or a thousandth of an Angstrom, but it's going to be pretty good. If I show you a simple sugar, you will be able to understand the most likely lowest energy conformation, and you will be able to generate some higher energy conformations. You can do that because you have models of cyclohexane and models of steroids, and models of simple ring compounds, and models of the forces of structural chemistry in your head. Then you utilize higher level reasoning to put these models together and to create the whole molecule from these rather small models. What we wanted to do was try and create a programme which followed this same line of expert reasoning. We would use simple models and combine them using some form of logic to get the final result. This can be done using a standard expert system, and that is what we started with.

A standard expert system is more or less shown on this slide; we have a fancy programme called the inference engine; it's just a simple, logic programme, which does very simple logic. We have a computer programmer, who calls himself a knowledge engineer. The basic difference is that he works with logic and he gets paid about ten thousand more a year. He talks to a human expert, and tries to get a set of rules, rules about the domain, rules of how to proceed with an analysis and so forth. These rules have to be somewhat general. They also have to be written in a fashion which is usable by this logic programme. We then present the case at hand. The case at hand is presented in a logical fashion as well. All of this together then allows the expert system to come up with a set of conclusions.

How did we structure WIZARD? Well, what we did was to build an expert system, which utilized the output from a molecular input programme, such as mimic. We did not try to make this thing a fancy graphics input and output programme. That has been done, I didn't want to reinvent the wheel. So, we are just took the input from some existing programme. We then wanted to generate a series of suggestions for likely conformations. Now, many times these estimated conformations are sufficient in and of themselves for the purpose at hand. However, if you need to, you can then subject them to quantum mechanics, molecular mechanics, or any further type of refinement programme. So, that is our expert system, and this is what I'm going to be talking about. We do not have any molecular graphics built directly into WIZARD.

We next needed to have the basic set of axioms to put into the logical system. The axioms are unsupported rules or statements about chemistry. They are feelings of belief, things we believe to be true. For example, in Euclidean geometry, we believe that two parallel lines never meet. Now, that seems pretty obvious, but in fact you can find that in certain Einsteinian spaces (which are curved) that parallel lines do meet. So, in fact we just have a series of unsupported facts which may have nothing to do with reality, it's all a mind game. We next have a series of premises which are observable facts. For example, a premise might be a molecule and how the molecule is connected together; the sorts of atoms and bonds inside of the molecule. We also need to have a logic. Well, we are using a very simple logic, the predicate calculus. It's been around since the last century and it's been revised by many logicians since then. These three things together form what is known as a model. The model is nothing but a series of statements which logically follow from the axioms and the premises. If this model predicts things that you see in the laboratory, then you say that the model created by the axioms seems to experimentally mirror reality. Some people go so far as to say that the axioms are true. Well, that's wrong. All you can really say with any axiomatic theory, is that the model created mirrors reality. That is true for Euclidean geometry, for quantum mechanics (which is nothing more than an axiomatic theory) and for the WoodwardHoffman rules. These are all very handy and useful axiomatic theories.

Why did I want to use axiomatic theories, which seem a bit more complex than standard programming techniques? Well first of all, we have seen that axiomatic systems such as Euclidean geometry, quantum mechanics, and such, are very useful. Secondly, the use of an axiomatic theory allows us to focus only on the premises and the axioms, and the logic. Logic is a pretty well founded theory. There are problems with logic, but the logic we use is very simple every day logic. Consequently, I don't have to worry about FORTRAN go tos, and loop statements, and input and output. I have complete freedom from those kinds of problems. All I have to do is make logical statements about chemistry and then use logic to combine them to get a model. So the use of an axiomatic system and logic programming frees me from a tremendous burden of programming responsibility. Also, (and this is tremendous fun) those individual axioms about chemistry can be manipulated by another programme. If you have written FORTRAN programmes you know how hard it is to manipulate a FORTRAN programme by yourself, much less to try and create another programme which can manipulate a FORTRAN programme. But logic statements (because they are so simple, and I'll show you what some of them look like) can easily be manipulated by another programme. This opens up such things as the ability to check for consistency, for completeness, for selfcontradiction, and to create learning systems. So axiomatic theories actually provide us with tremendous strength and flexibility at a cost of a little bit of certain powers. (For example, trying to create an axiomatic theory of molecular graphics would be silly. It would take too long to prove a theorem, that would put a molecule up on a graphics screen). Axiomatic theories don't do everything, but they are wonderful.

Now that we have defined what an axiomatic theory is, and how we are going to use it, you need to know how we got our axioms; where did they generate from; and what are they? Well, the first thing I did, (going back to the expert system) was to grab several experts. I sat in a room with them, took in six packs of beer (so that they thought it was worth their time) and talked to them. Out of that, we came up with a very simple basic rule of how to perform a conformational analysis. They said that certain connected patterns, of atoms and bonds behave as a unit. It can be recognized as such and treated as such. If I were to draw a structure of ditbutyl cyclohexane, you would recognize cyclohexane and you would recognize tbutyl groups. You would not try and analyze each bond in the cyclohexane independently, because you have knowledge of the behavior of cyclohexane. What we do in WIZARD is to recognize those kinds of natural units that behave in a special fashion. We then recognize that these natural units have a limited number of behaviors. They are not totally unpredictable, otherwise chemists wouldn't identify them as being something useful to recognize. Take for example, cyclohexane; we recognize that as a useful thing to think about due to the fact that there is a chair, a boat, a halfchair, a twistboat, i.e. a limited number of natural subconformations. We then take those natural units, assign them subconformations and put them together. Basically, what we are doing is to build a Dreiding model inside a machine, nothing more fancy than that. While we are actually building the model, we check to make sure we are not doing anything stupid along the way. For example, consider trying to build cholestane from a series of boat cyclohexanes; if you did it in one way, it would curve back on itself and the D ring would run into the A ring. So we just have to make sure that we don't do anything stupid. If we have done something that gives us a problem, we have to resolve that problem. Either we can ignore it, because there is another suggestion that doesn't have a problem, or we can try and fix it. This is the basic rule, the basic axiom of conformational analysis that we work with. In a bit we will see what the results are that come out of this.

First let us look at an example of a unit, the cyclohexane unit. We have a pattern that allows us to map it on to the molecule. This pattern consists of atom types, bond types, and stereochemistry. Underneath that is a list of what we call templates, the natural subconformations, the three dimensional coordinates of the thing. That's given by the XYZ coordinates of the templates. Here is the chair template; we'll also have a boat template, a twistboat template (in the plus and minus form) a halfchair template, a flat template, and other templates. For each template, we store certain pieces of abstract information. For example, the torsion angles around the edge of a ring. We store those as; staggered, nearly staggered, nearly eclipsed, or eclipsed. These are all obviously staggered. We store information about symmetry and we store information about the flexibility of the system. All this information can then be utilized in later reasoning processes.

Where do we get this information? Well, we need to go to experts or to data to find it. We can take a molecule which exemplifies a type of unit that we are interested in. We can rotate it using molecular mechanics and get this sort of curve, find the natural minimum energy conformation, and store that, along with information on flexibility and other things. How do we store flexibility? We say, for example, that this is a minimum energy, and then we say for a twist of plus ten degrees, it costs so many tenths of a Kcal, for twenty degrees, so many tenths of a Kcal, and so on. We do not store any complex mathematical functions, we store simple tables. There is a couple of reasons for that. Prolog is not a language which is really well suited for doing math more complex than a simple squareroot, or a distance function. The other thing is that I'm pretty simple myself and I like to try and do things by addition and subtraction because I can understand that. Thus we just simply store tables of values.

In certain places we cannot get information from molecular mechanics because we do not have parameters (for example this kind of cobalt case). In that case we store information about the basic structure, but we do not store any flexibility information. So WIZARD uses as much information as it can get. When it needs information that it doesn't have (for example, if it is asked; "What would the flexibility of this sort of thing be"?) it would come back and say, "I'm sorry, I don't have enough information to answer that question", but it doesn't crash. This somewhat mirrors expert behavior. It does not give you a bad answer when it doesn't know how to calculate the answer; it tells you straight out, "I'm sorry, I'm ignorant in this case". So, anyway, that is how we got the information to utilize in the unit frames, the knowledge which is necessary to assign the basic on conformations to the units.

Let's analyze a real molecule, and I'll show you how it's done. The first step you need to do (if you remember in our basic rules) is to find the units. Now I think everyone who would want to try and do a conformational analysis of cyclazocine would do the same thing. You would recognize the different bits in it and say; "well these are the kind of things that I have to understand the subconformations of so that I can understand the conformation of the whole thing". There is nothing surprising about this. WIZARD does this; it recognizes nine units as comprising the cyclazocine molecule. It then puts these back together in a hierarchical graph. The lowest description is the actual atom and bond connectors. The next description talks about the fact that unit one is connected to unit two by a bridge fusion, and unit two is connected to unit three by a ring fusion. This allows us to perform certain types of reasoning. We know, for example, that two acyclic units that are joined by an acyclic bond are likely to have free rotation. We know that two rings that are joined by a fusion are still likely to be flexible, especially if the original rings are flexible. We know that bridged systems are likely to lose flexibility. This sort of graph is something that WIZARD will utilize to reason about when a molecule is likely to lose or gain flexibility.

WIZARD goes even further than this; it creates an even simpler graph. This more abstract graph says that here is the core extended cycle of cyclazocine. It realizes that this sort of extended cycle consisting of bridged and fused rings all put together is likely to be the most constrained parts of the molecule. Consequently it tries to build this first. Then, after it has built the central part of cyclazocine, it tries to build on the outward parts. This is done to make the programme more efficient.

WIZARD also has another way of being efficient, it tries to reason in an abstract way before it reasons in a detailed fashion. I think I can show you that a human expert does the same thing. Let us take the example of cholestane. If you were to try and build a model of cholestane out of boat, boat, boat, envelope, A B C D rings, you can reason in your mind that if you join up the boats so that the bow touches the stern of another boat you get a kind of a deep cup. As you put another boat cyclohexane on that, you have made the thing even more deep. You haven't done any reasoning with XYZ coordinates, you have not utilized any geometry beyond an abstract geometry, and yet you can realize that trying to construct cholestane out of these kind of rings will lead you to problems if you do it stern to bow, stern to bow. So you recognize at that point that there may be a problem here. WIZARD tries to do the same thing. It works at these various levels of abstraction (which I showed you before) to try and determine whether or not we can predict ahead of time that there is certain to be a problem. Then because of the nature of the one to many mapping we use (many conformations map to one abstract conformation, several abstract conformations may go to one higher in the tree) we can show that if an idea is stupid at the most abstract representation then we can guarantee that everything that would actually be a detailed representation following from that would also be silly. Consequently this is the abstraction technique we utilize.

Now to give you an example of how that actually works. In cyclazocine we have nine units. The programme makes an abstract suggestion. We are not going to be doing any geometry at this point, we are only going to be working with symbols. WIZARD says, "let's put unit one (which is a cyclohexane) into a chair conformation" (these b's are symmetry information, ignore them). Cyclohexane number two is going to go into a halfchair. We assign symbolic conformation names to the things, we do not assign coordinates yet. Then we try and manipulate these things symbolically. For example, if you remember I had the symbolic torsion angles staggered, nearly staggered, eclipsed, and nearly eclipsed, in the unit knowledge frame. We know that if we are trying to join a chair and a boat together, and if we try and join a staggered bond to a staggered bond that works quite well. If we try and join a staggered bond to an eclipsed bond we will find that the overlap will not be good, that we would have a tremendously strained atom, and we can't do that. That is another example of reasoning at an abstract level; we are not doing any geometry, we're not doing any molecular mechanics, and yet we can say that something built in this fashion is a bad idea, while something built in another fashion is possibly an acceptable idea. We try and do as much reasoning as possible at that level. We try and compare the torsion angles on the edges, and we try and check to see whether or not we're trying to do something which is a well known "bad case". For example, we might be trying to violate the pentane rule. (I'll show you an exact example of the pentane rule later). Or, perhaps we have already tried to build a different conformer and we learned when actually utilizing XYZ coordinates that it was a bad idea. These are all ways that we can try and criticize the thing at abstract levels.

Eventually we get something which is not criticized at an abstract level, something which looks good without actually using geometry. At that point we have to take the XYZ coordinate templates and put them together. If we're building the basic ring skeleton of cocaine, (as shown in this slide), we would take the chair cyclohexane, the envelope cyclopentane and we try and put them together so as to minimize the angular deviation between the two. With cyclazocine, this is the sort of way we try to build it up from the templates. We would take knowledge of our individual subconformations, we would put them together in an abstract fashion, we would criticize that in an abstract fashion, then we would then actually build it to come up with a few suggestions that got all the way through the criticism.

Now, occasionally when you are actually putting cyclazocine together you would perhaps find that a hydrogen here would run into a hydrogen over there. We can recognize that by calculating cartesian distances. We are still not using molecular mechanics or anything more fancy that adding up van der Waal's radii. If we see that a certain suggestion is a bad idea, we would store that and utilize that in an abstract fashion by saying, "if we suggest that this is a certain subconformation of the whole and we've already tried that, and found it to be bad, then don't try it again". That is another way of getting abstract criticism from details. But eventually something is not criticized at the abstract level, and it is not criticized while it's being built up, and it actually is created. An example of such output is shown in the next slide.

This is WIZARD'S final suggestion for cyclazocine, overlaid on the xray crystal structure of cyclazocine. Remember that we did not use any molecular mechanics or quantum mechanics. We didn't utilize any numerical reasoning other than trying to minimize angular overlap when we put the templates together and calculated distances. I would say that that's a pretty good fit. Other examples of how well WIZARD does on putting molecules together is shown here. This is a slightly more complex compound, it's like cyclazocine but it has another six membered ring fused to the core. The next slide shows the results for dichlorocholestane. Basically I would say that we are doing pretty good at calculating crystal structure using logic.

There _are_ certain things that we cannot do, yet. We do not make a good estimate of energies. We do a pretty good job of guessing crystal structures, but if you want energies you have to actually minimize it. That is due to the fact, as someone said yesterday, that the energetic contribution consists of not only the gross factors but all of the little factors put together, things that you might be tempted to ignore. In fact, the van der Waal's contribution between these distant hydrogens (when you get enough of them) can actually add up to a few Kcal, or so.

Not only can we predict crystal structures, but we can also be useful to chemists in the laboratory. Laurence Harwood, at Oxford, was looking at this intramolecular DielsAlder reaction. He asked us for help in trying to understand this reaction, because it was not a simple classical DielsAlder. He was getting mixtures of endo and exo products, and these mixtures were changing with time. So, obviously we are seeing both a thermodynamic and a kinetic product being formed. What we did with WIZARD was to give it a distance constraint based on the fact that the dienophile had to lay within 3.5 A of the diene. I was very happy yesterday to hear the number 2.2 A (as a DielsAlder transition state distance) because that means we easily overestimated that distance. We found twentyeight conformations in this case. Then we analyzed all the lower energy conformations in the exo and endo products. Then we tried to figure out which conformation went to which product. We did that by looking at a minimization of angular strain and of distance. For example, here is the product, and here is the starting material, and by looking at this sidechain it's pretty obvious that this starting material leads to that product.

Once we had identified a correlation between starting materials and products, we made a correlation diagram in which we took each starting material and linked it to the corresponding product, and then we found some interesting and exciting results. For example; in this case even though the endo is the most stable conformational product, it comes from a very highenergy starting material. The lowest energy starting conformation goes to a exo product. I can also give an endo product, but look at the difference in energy of the transition states. It's going to be a very slow reaction going to the exo. Yet once it goes to the exo, since the reaction is reversible, we will tend to see a slow conversion to the endo. The exo, (which is the dotted line) will slowly drain over to the solid line, the endo product. By having the complete set of conformations available, and by correlating them in this fashion, we are able to demonstrate the relationships between the thermodynamics and kinetics in this series.

Since we were working with approximately twenty examples of this, we calibrated ourselves on twelve examples, and then made predictions on seven. Actually wrote them down and gave a copy to

Laurence Harwood (and I was very worried because my reputation was on the line). Out of those seven predictions, six of them were correct and on the last of them the experiment just didn't work (so he's trying to remake the starting material). I think that argues pretty well that WIZARD has done a good job of helping to understand this reaction.

What's the cost and benefit of using WIZARD, rather than just trying to do it with standard, classical bond angle driving? Well WIZARD running on a VAX 11/750 (which is a fairly slow machine), took something like one hundred and fiftythree CPU seconds to do the whole analysis for this reaction product. MM2 took five hundred and fiftythree seconds per conformation to minimize each one of these (remember that a VAX 11/750 is a slow machine). So the total combination of nine conformations times five hundred and fiftythree plus the WIZARD time was five thousand one hundred and thirty CPU seconds. The cost of using WIZARD as a filter was approximately three tenths of one minimization. WIZARD suggested two conformations which turned out to be actually degenerate, they were just a little to each side of a minimum. So we'll call those two duplicates that were shown up by the minimization as an error. So the total WIZARD overhead for utilizing a combination of WIZARD and MM2, was the same amount of time as 1.3 minimizations, (actual CPU time plus time wasted on a duplicate).

Next let us consider how difficult it is to do the same study by driving torsion angles. It is well known that trying to find all minimum energy conformations in medium rings is a difficult task. I do not know exactly how small of a angular step we would want to use, but I am sure that if we wanted to be exhaustive we would need to do quite a few minimizations that yielded duplicate structures, certainly more than 1.3! So the combination of using an expert to make suggestions, and then a molecular mechanics programme to refine those suggestions, is substantially faster than simply using an angle driver to rotate all of the bonds. I don't think anybody would be too surprised to find that utilizing a bit of intelligence before you use brute force is a good idea. We saw that demonstrated in Doctor Kitson's talk yesterday, in which, instead of just using the Cray XMP without being careful, they first used intelligence to relax one bit and then another bit and another bit and finally relaxed the whole thing. If you don't have infinite CPU time you have to use intelligence. All we are doing is providing a bit of machine intelligence here.

We do have certain problems that arise though. For example, this was WIZARD'S _initial_ suggestion of cocaine; it's not bad, but we noticed the sidechain is badly off. If you remember the basic rule, I said we have to examine each suggestion for problems. This has a problem in that we have an unbalanced electrostatic attraction between an electropositive nitrogen and oxygen lone pairs. This is WIZARD'S suggestion here, with the oxygen out here, while the crystal shows the oxygen over here. The reason for that is that WIZARD wasn't taking into account the electrostatic attraction in the initial idea. So here is a problem; WIZARD can look at this and say, "I haven't taken into account the electrostatic attraction, I had better do that." How is it going to do that? Well, let's go way back to the saturated hydrocarbons. Remember, this is what I said, "we first see if we find a problem, and if there are problems we try and fix them". In saturated hydrocarbons the classical problem is where two hydrogens run into each other. This is known as the pentane violation. Now, we took this case and subjected it to MM2 analysis, and we found out that one bond or the other bond would twist, but not both bonds. This kind of surprised me, and so we went to the Cambridge Structural Data Base, and looked to see if that was true. We found approximately a hundred and seventy compounds that had this sort of structure built into it. Over one hundred and fifty of those compounds showed only one bond or the other bond twisting. I personally would have thought that it would have been a symmetrical twist, but that turns out not to be the case (if anybody wants to know why, I can tell you of the month I spent trying to figure it out, and the absolutely simple and disappointing answer why it's this way and not the symmetrical spread).

Anyway, so what we did way back when we were just doing saturated hydrocarbons was to create a rule. This is where expert systems start to fall down, and where logic programmes don't fall down so easily. With an expert system when you find a case like this you create a rule. In this case the rule said; "when we have a pentane violation in the molecule, twist one bond or the other bond." Well that worked pretty fine in this sort of case, for example, when we looked at the bond rotation around this we found that either one bond or the other bond could twist, forming a little doublet (kind of like NMR splitting). So I call this conformational splitting. When you add another methyl you get another little doublet. You add another methyl you get another little doublet. That's exactly (if we go back to the previous slide) what we are seeing; one bond twists or the other bond twists and you get a pair of neighboring conformations. That by the way, is a take home lesson for your people who are doing minimization with MM2 or similar programmes. If you have pentane violations and you analyze this with minimization once and only once, you'll get either this compound or that compound, but you'll miss its little brother which is sitting just a few degrees away. So be very careful when working with strained compounds, they tend to have very exciting behavior.

But to return to WIZARD, we made up this rule and we added it into our simple expert system. That expert system did very well with these sort of things, but I had problems with these sort of things. This doesn't twist, it bond opens. The reason for that is that if you try and twist any single bond, certainly you relieve the strain between the two forward methyl groups, and you relieve the strain between the two backward methyl groups, but you introduce a cross term which is worse than either one. Well, the answer to that in expert system technology is very simple; add another rule that says; "if you have this sort of problem you just spread the angle rather than trying to twist it". So we started adding rules for saturated hydrocarbons, and we kept getting further and further and then we would have a problem and add a rule, then we would get further and further and we would have another problem and have to add another rule, etc.

For example, the bond spreading rule worked fine until I put in molecules with threefold rotational axis. In that case it would open up the angle in the front, then it would open up this angle, and then it would open up that angle, which closes the first angle. And it would go around the circle, go around and around. So I just simply added another rule, and then I had to add another rule, and another rule. And finally I realized that instead of creating artificial intelligence, I was creating <u>artificial stupidity</u>, because the programme wasn't thinking about what it was doing, it was just doing it. There was no reasoning, there was no intelligence, it was just doing a 'lookup'. This is one of my arguments against expert systems. They do not reason, they only look things up. So what are we going to do?

There is another problem; if you are going to have rules about every case that is bad, then every time you add a new bond type, you will have to add new rules for that bond type. The problem here is not the fact that we have a carbonyl, two single bonds, a double bond, and a hydrogen, per se. That list of bond types and atoms is not the problem. The problem is that this hydrogen is running into that oxygen. So instead of adding all of these special case rules and saying, "when you see this do this or that". What we decided was to look at the root cause. In this case the cause is that there is van der Waal's strain. Now that we have identified the root cause, let's try and solve the root cause. That's reasoning, that's not rule lookup.

So we took the standard expert system which starts with premises and works up to a conclusion and we allowed it to look backwards at the same time. For example, if you create a molecule which has a van der Waal's problem by a chain of reasoning, then that chain of reasoning cannot be correct. So instead of having a very complex chain of reasoning that does everything we break it up into a couple of simple steps. We make a suggestion and then criticize that suggestion based on such things as van der Waal's information. If you come up with a criticism of that suggestion then you know your chain of reasoning was incorrect, and then, and only then, do you examine the molecule in more detail. So we added critics about van der Waal's forces, so that when we made a naive suggestion it could look to see whether or not hydrogens were banging into each other. We added critics about electrostatic forces, so that in the case of cocaine we could recognize that the oxygen and the nitrogen would like to be closer together and so that we could say, "let's see if there is a way to pull them together." We added critics that allowed the programme to reason better; for example, if the programme tried to twist a bond in the plus and minus fashion, you don't have to have a special case to tell you that you cannot _do_ something and _undo_ it simultaneously and get anything useful out of it. Then we made a simple stepwise method; when you find a criticism in the molecule, what you try to do is to twist angles, and if the twisting of angles doesn't work then, and only then, try to go to bond angle changes. If you cannot solve the problems of the molecule with bond angle changes then you try and do bond length changes. So, first we look for chemical problems, then we try to find a chemical way to solve them, we implement that through geometry and we make sure that during the geometry we are not trying to do anything stupid. This is more of a reasoning process that a lookup process.

Let me show you how that works with ditbutyl methane. For example, we now reason that this is a problem. We don't reason that it is a problem because we have a rule that says; "if we have a gauche plus and a gauche minus bond that's a problem", we reason it's a problem because we add up the van der Waal's radii and find that they are too close. Then having reasoned that there are two problems we try and solve each problem simultaneously. What we want to do is try and exclude this hydrogen van der Waal's radius from that one, and at the same time we want to try and exclude this hydrogen van der Waal's radius from that one. So our first simple suggestion without any special case knowledge, without _any_ knowledge of this sort of case at all, is to just try and twist bond number one by plus twenty degrees. To solve the other criticism in the back, we try and twist bond one by minus twenty degrees. Our arithmetic critic now comes up and says; "Wait a second, you are being stupid. You cannot twist a bond by plus twenty degrees and minus twenty degrees, that doesn't work". So next we try twisting bond one by twenty degrees and bond two by minus twenty degrees, that's a very large twist, and the programme now recognizes that we have a chemical problem, rather than a geometrical problem. Any set of angle twists on this sort of system, will give us either an arithmetic problem or a chemical problem.

So after having reasoned at its first level of energy, the programme says; "I cannot do anything with just simple bond twisting, I have to go to angle opening". So it now does its first simplest angle opening, it tries to open that angle by fifteen degrees. The geometric critics say, "that's fine," and the arithmetics critics say, "that's fine", the chemistry critics say, "that's fine". So WIZARD, now having tried that level of chemistry says, "I cannot see anything wrong with it, let's suggest it." That turned out to be the structure of ditbutyl methane according to an electronic diffraction, the central bond angle opens by about 17.2 degrees.

Now, let's look at that reasoning process and how that reasoning process works with some strained molecules. For example, flufanamic acid would like to be planer due to the fact that this analinyl wants to be planer and this analinyl wants to be planer, but we simply reason that these two hydrogens bang into each other.

So the chemistry critics say we have a problem. We have never seen
this case before, we are not following a complex rule example like
an expert system would. We have reasoned that there should be a
problem by a very simple chemical axiom, van der Waal's critic.
So, now the programme says, "now that we have got a problem how are
we going to deal with this problem?" Well it's going to try and
twist bonds, but it was to select the bond to twist. If you
remember I put in bond twisting information by little tables. So
let's go back to flufanamic acid and see what it would do. It
would use very simple geometry and reason that if you open this
angle up by fifteen degrees you will separate these hydrogens
enough. It also says that if you twist a ring bond you have to
open that up by almost forty degrees to get enough movement. It
looks into the frames, and it finds that, twisting one of these by
forty degrees costs a lot of energy, tens, twenties, of Kcals,
whereas twisting one of these acyclic bonds by fifteen degrees,
doesn't cause many Kcals. So it goes through a very simple
reasoning process to decide which bonds to twist. Now for
flufanamic acid, there are two known crystal morphs. WIZARD
suggested that there should be four low energy conformations. Here
is one of WIZARD'S suggestions overlaid on one of the known crystal
morphs. We haven't done any molecular mechanics or quantum
mechanics, all we have done is take existing templates, make a
simple suggestion, find that there is a problem, (these hydrogens
are banging into each other) and simply calculate how far you have
to rotate one of these bonds to get the van der Waal's radii so
that they just kiss rather than overlap. Here is the other one of
WIZARD'S suggestion over the other known crystal morph. In
cocaine, if you remember, we wanted to bring the side chain around
so that this oxygen is pointing its lone pair towards that
nitrogen. All we did was try and get it so that the van der Waal's
radius of this lone pair and the hydrogen here touch. WIZARD
looked at the various bonds in this system and found that by
twisting around this bond you would get the most change in the
position of this oxygen for the least amount of energy. So it
suggested rotating this bond by something like fortyseven degrees
and the final suggestion made by WIZARD after that is shown here,
overlaid on the crystal structure of cocaine. This next compound
is a [3.3.1] bicycloamino acid. If both of these two rings were
in pure chair forms, these hydrocarbons would come within .57
Angstrom of each other, which is far too close, so WIZARD says, "I
have to do something to get these things apart". What it does is
to suggest that this ring be twisted by twenty degrees in the
positive fashion, which forces a negative twist of twenty degrees
in this ring so that they overlap correctly. It didn't suggest
that this complimentary twist be done because of symmetry. It
looks like it has to be due to symmetry but in fact that occurs due
to the fact that if you twist this ring by twenty degrees, the
resulting bonds coming off of it have to be twisted by twenty
degrees.

That sort of symmetry thing, where simple reasoning causes symmetry to be taken into account is very fun. For example, let us look at triisopropyl methane. If we look at the possibilities of twisting, we find we could twist this in a positive fashion and we could twist that one in a negative fashion etc., etc. If we do anything except twist all these in a negative fashion, we end up with hydrogens bumping into each other. So the fact that what happens to this molecule retains a C3 axis of symmetry is just simple chemistry and simple geometry, and yet this is the principle of corotation which has been put forth by Mislow, and studied by many people. Corotation falls out of very simple geometry.

Let's look at the success of our attempts to find all conformers. So far we have done pretty good. We have challenged WIZARD'S results by looking at it with MonteCarlo methods and a few other methods. We have sometimes found that WIZARD has made excess suggestions. It was suggested that something might be a stable conformation when in fact it was just a little shoulder on a slope. But we haven't missed any conformations yet. <u>Yet.</u> I'll admit, we are going to fail someday. The suggestions are good. I think having shown you the xray overlays that you can agree with me that they are decent. The process is efficient. Its one hundred or a thousand times more efficient to use the combination of WIZARD and molecular mechanics than utilizing MM2 alone. If you throw out the molecular mechanics, if all you want is structure rather than structure energy, we can go up to ten to the fifth times more efficient that utilizing molecular mechanics alone. We can also provide answers even where numerical programmes cannot. We can utilize data taken out of the structural data base to look at systems were there are no parameters, so that we can do things with cobalt or titanium complexes where you cannot utilize a molecular mechanics programme. There is a student at Oxford who is looking at that right now.

So to summarize, to date we have created a useful tool for chemists. The studies with Laurence Harwood were really useful in understanding that system. We better understand the axioms of conformational analysis, we know that certain things that we thought were important, are not. Other things that we thought were trivial turned out to be very important. So I think after having done this, I understand the process of conformational analysis a lot better. We are currently working on looking into new classes of compounds and to improve the efficiency of WIZARD by better planning the attack on any individual molecule. Future directions are; looking at the axioms themselves, reasoning about the axioms, checking to make sure that they are consistent, and checking to make sure that they are all there. Trying to learn from things like the Cambridge Structural Data Base.

I would like to thank a number of people. I would like to thank Dr. R.E. Carter, who allowed me the freedom to start my own project as a postdoc in his group. I would like to thank Dr. C.K. Prout, who allowed me to continue this project at the Chemical

Crystallography Lab at Oxford University. I would also like to thank in addition to those two people, Mr. Andrew Leach, who is the graduate student at Oxford who did a lot of this work. He did all the hard geometrical reasoning and such. (I'm hopeless with geometry). I would like to thank all of my experts, and then I would like to thank you for allowing me to have this chance to talk.

STATE OF THE ART IN VIBRATIONAL DYNAMICS OF LARGE MOLECULES

Giuseppe Zerbi

Dipartimento di Chimica Industriale, Politecnico, Milano.

1. INTRODUCTION

In this paper we wish to review the present state of the art of the computational aspects in molecular dynamics and vibrational spectroscopy. We wish to summarize in a schematic way the present computational facilities offered to a scientist who wishes to tackle the problem of molecular and lattice dynamics.

The references in this short review article will be few; the references quoted, however, are not single specialized articles, but books or review articles where the interested reader can find a complete discussion of the concepts and results mentioned in this paper.

First a clear definition of the frequency domain which the term "molecular dynamics" in our case refers to. When the amount kT of energy is provided to a macromolecular system, molecules can perform localized and collective large amplitude motions whose displacements can or cannot be recovered elastically depending on the frequencies, rate of motions and amplitudes. Collective "brownian" motions of macromolecules and polymers which determine the viscoelastic and mechanical properties of organic polymeric materials can occur in the frequency range from 10^{-24} Hz to megahertz. The mechanical relaxation properties of these materials are related to the onset of specific motions of chain segments or of side groups; the onset of these motions opens preferential channels at the molecular level through which the energy provided to the system during a periodic mechanical perturbation can be dissipated. Non recoverable strain may occur. The frequency range of 10^{-4}-10^{6} Hz can be studied with suitable mechanical systems and NMR.

The kind of molecular dynamics we discuss in this paper refers to mostly elastic infinitesimal vibrational displacements of atoms about the equilibrium configuration at frequencies of the order of 10^{12}-10^{15} Hz. It has been shown that vibrational displacements can be highly topologically and energetically localized or can have a cooperative or collective character. Vibrational normal modes are thus defined for a finite molecule. If the system has one- or tri-dimensional periodicity phonons are

created.
These motions may be driven resonantly by an electromagnetic wave with energy in the infrared when dipole allowed transitions may occur or may scatter coherently or incoherently a beam of thermal neutrons. Optical infrared and Raman or neutron spectroscopy are thus the techniques presently used for the study of these motions.

The detailed knowledge of the dynamics of molecules provides a way to assign the very many absorption bands or scattering lines measured in spectroscopic experiments to specific motions of groups of atoms. One can then describe the structure of the molecule and its behaviour in various kinds of chemical environments or in various physical situations.

2. THE MOLECULAR MODEL AND ITS DYNAMICS

Most of the classical molecular dynamics treatments consider a molecule as a set of balls connected by weightless springs [1,2]. In other words we content ourselves in considering a N atomic molecule as a set of 3N-6 coupled harmonic oscillators which perform 3N-6 normal modes of vibration Q_i in which all atoms move in phase with frequency ν_i (i=1,3N-6) and with amplitudes L_i weighted by their masses. The choice of hookeian springs between atoms implies that the restoring force be linear with the displacement, $f_x = -k_x x$, through the force constant k_x of a harmonic potential. The potential energy can be written for a single oscillator as:

$$2 V = k_x x^2 \qquad (1)$$

and for the whole system in matrix notation as:

$$2 V = x \, \tilde{F}^x \, x \qquad (2)$$

where x is the vector of the cartesian displacement coordinates and F^x is the matrix of the quadratic force constants:

$$f_{ij} = (\partial^2 V / \partial x_i \, \partial x_j)_0 \qquad (3)$$

Contrary to the basic concept of chemistry, a quadratic potential would imply that each bond cannot be dissociated. The approximation accepted by classical molecular dynamics derives from the fact that the generally unknown molecular potential is expanded in a Taylor series about the equilibrium geometry for infinitesimal vibrational displacements. And indeed the amplitudes of any normal mode about the equilibrium position in any molecular systems are extremely small. Higher order terms can be and have been included in the detailed treatments of the vibrations of a few small molecules; the discussion of higher order terms is outside the scope of this paper.

Vibrational molecular spectroscopy has enjoyed a rapid development in the theoretical aspects when the chemical concept of bonds has been introduced and new types of coordinates have been defined. Let $R = Bx$ define a set of internal displacement coordinates as linear combinations of cartesian displacement coordinates x. The R's describe bond stretching, angle bending and torsions.

The vibrational potential energy of Eq. (2) can be rewritten in terms of R's as:

$$2V = \tilde{R} F R \quad (4)$$

and the kinetic energy can be written as (see ref. 1):

$$2T = \tilde{R} G^{-1} R \quad (5)$$

where the matrix G contains information on atomic masses and molecular geometry.

We wish to find the linear transformation Λ from R's to Q's such as:

$$2T = \tilde{\dot{Q}} \tilde{L} G^{-1} L \dot{Q} = \tilde{\dot{Q}} E \dot{Q}$$

$$2V = \tilde{Q} \Lambda Q \quad (6)$$

The basic concept of normal modes requires that:

$$\tilde{L} F L = \Lambda \quad \text{and} \quad \tilde{L} G^{-1} L = E \quad (7)$$

which brings to the eigenvalue equation:

$$G F L = L \Lambda \quad (8)$$

which is the basic equation to be solved in any dynamical treatment of molecules. In Eq. (8) Λ is a diagonal matrix with $\lambda_i = 4\pi^2 c^2 \nu_i^2$, where ν_i is the vibrational frequency in wavenumbers. Eq. (8) can be rewritten as:

$$\left| G F - E \lambda \right| L = 0 \quad (9)$$

which shows that non trivial solutions can be obtained by finding the roots λ of a polynomial equation of $3N-6$ order. The matrix of the vibrational displacements can be found with Eq. (8) once Eq. (9) is solved.

3. THE VIBRATIONAL POTENTIAL

Actual calculations can be carried out when the matrices G and F are numerically known. The elements of G are usually known since precise geometrical parameters have been measured on small molecules by several independent techniques. Parameters for large molecules can be safely transferred from small model molecules.

The main difficulty lies in the determination of the numerical values of the elements of F for a valence force field (VFF). This problem is not fully solved, but at present sets of highly reliable internal VFF parameters are available for the most common classes of organic molecules.

The origin of these numbers is the following:
1) In the past twenty years accurate spectroscopic studies on classes of molecules have provided sets of reliable vibrational assignments of infrared and Raman spectra [2-4]. Since the number of parameters is always larger than the number of experimental data, frequencies and vibrational assignment of very many isotopic derivatives (D, N^{15}, C^{13}, etc.) have been collected. Since, within the Born-Oppenheimer approximation, the vibrational potential is isotopically invariant, the experimental data can be used together with the data from the parent molecules to calculate the elements of F by least squares refinement. Since the experimental data are still not enough, chemical common sense has suggested to use chemically similar molecules which may have similar potential. Moreover, other data from independent physical phenomena have been used in a grand least squares refinement. Among the additional data we quote i) Coriolis coupling coefficients and centrifugal distortion constants for gaseous molecules, ii) Debye Waller factor (or thermal factors) in electron, X-ray and neutron scattering experiments, iii) specific heat and elastic moduli in solid samples, etc. and iv) more recently vibrational infrared and Raman intensities.

Reliable sets of VFF force constants are presently available for large classes of organic molecules and the reader is referred to specialized publications [2-5].

2) More recently a great help in force constant calculation has been derived from quantum mechanical calculations. The availability of very large and fast computers has allowed to perform semiempirical and/or "ab initio" calculations with large basis sets. Many molecular properties can then be predicted and when suitably scaled they can be used in the calculation of the vibrational spectra.

In this paper no mention is made on the concepts of "molecular dynamical models" which have been successfully applied in conformational analysis of large molecules. This problem is discussed in other papers of this book. As usual, groups of specialists in one field are unaware what are the problems, the achievements and limitations of other techniques used by other groups in parallel problems of molecular physics. This has happened also between the world of spectroscopy and that of conformational analysis.

Spectroscopists like to consider, in a very comprehensive way, the properties of a given molecular force field. First of all generally a VFF includes forces between covalent bonds (as chemistry teaches) while often in conformational analysis (MFF) interactions between non bonded atoms play the most relevant role. Moreover a VFF should reproduce all the vibro-rotational constants derived from a rovibronic analysis of molecules in the gas phase. In a comparison between results from VFF and MFF the fitting with experimental frequencies should be made on small molecules after symmetry factoring has been applied. A comparison with spectra of large molecules is not feasible since vibrational assignments are extremely uncertain and vague in the case of large molecules. Indeed a band can always be found in a spectrum of a complex molecule to fulfill the requirements by a theoretical calculation.

4. COMPUTING PROGRAMS FOR THE CALCULATION OF NORMAL COORDINATES AND FORCE CONSTANTS OF SMALL AND FINITE SYSTEMS

Routine computing programs are available throughout the world for automatic and fast calculation of normal vibrations and force constants. Every laboratory of spectroscopy has acquired from some non commercial source some programs which have been later modified in order to suit a given kind of computer. In discussing the properties of these programs we feel however necessary to put in the right historical perspective the development of such programs. The equations which formed the basis for all numerical programs have been developed starting in 1958 in the school of Professor Bryce Crawford at the University of Minnesota. The perturbation of the eigenvalue equation [Eq. (8)] in terms of small changes in G, F, L and Λ was developed in a thesis by King in 1956 and published too late in 1960. The least square refinement of the force constants based on King's thesis was first developed by Curtiss in 1958 and never published. The first complete computing program became available in 1961, written by Drs. Schachtschneider and Snyder (S-S) (see refs. 4 and 5 for a list of references). These authors discuss in great detail all the necessary physics and mathematics involved in calculations of vibrational dynamics. Programmes were collected in technical reports from Shell Development Co. and were generously distributed in the world. Most of the programs presently available even with different names reproduce, often with no acknowledgments, the original programs by S-S. A fully independent set of more complete and more automatic programs has been later developed by the school of Professor Shimanouchi in Japan. The school of Dr. Snyder in California has at present added more flexibility and power to molecular dynamical calculations.

The input of these computing programs consists of the description of the molecule in terms of structure and atomic masses and the choice of the force field. The matrices B, G, F, L and Λ are calculated. A least squares refinement with observed frequencies can be automatically performed. As final output all the known vibrorotational constants are obbtained.

Symmetry factoring of the secular equation is generally achieved by providing the computer with the transformation coefficients for a given symmetry. To our knowledge none of the commonly used programs considers an automatic analysis of the symmetry properties of the molecule once the cartesian coordinates of the atoms are calculated. We have developed algorithms and applied in a computer for the automatic study of the symmetry of the molecular system. These are particularly useful when many degenerate species and many redundancies occur in the calculations (e.g. the calculation of normal modes of adamantane).

The size of the secular equation to be solved has always limited the kind and size of molecules to be studied. At present secular equation with 150 roots can be routinely solved with great numerical accuracy both in the eigenvalues and eigenvectors.

In the programs distributed by S-S, cartesian atomic displacements as well as all vibrorotational constants can be obtained as output. No physical quantities associated to tridimensional crystalline order (sublimation energy, Debye Waller factor, etc.) are given since the molecules are considered in their "gas phase", i.e. as isolated entities.

5. COMPUTING PROGRAMS FOR THE CALCULATIONS OF ONE- AND TRI-DIMENSIONAL LATTICES

Let us first consider a perfect regular one-dimensional array of chemical units linked by some sort of bond. If the molecule is truly isolated in space or if intermolecular forces are extremely weak as compared with intramolecular forces, one can deal with a perfect one-dimensional crystal. The translational periodicity of the lattice introduces the concept of the k vector and allows to rewrite the secular equation in the following form [4,5]:

$$G(k) \ F(k) \ L(k) = L(k) \ \Lambda(k) \tag{10}$$

If N is the number of atoms in the translational repeat unit the order of Eq. (10) is 3N and 3N frequencies can be calculated for each value of k. The calculated frequencies can be collected in 3N phonon branches of which in general 3N-3 are called optical and 3 acoustical branches. The corresponding vibrational displacements describe periodic waves which propagate along the polymer chains. In solid state physics these waves are called phonons. In lattice dynamical calculations the periodicity of the lattice allows to define a Brillouin Zone (BZ). For 1-D crystals the treatment of the BZ in the calculations is very easy.

Phonon dispersion curves have been calculated for several polymeric organic materials for the interpretation of the optical and neutron spectra [5,6]. Sets of automatic computing programs for the calculation of phonon dispersion curves and the derived one-phonon or two-phonon density of states have been developed in our group in Milano and applied in many practical cases.

The calculation of phonon frequencies and of density of states for 3-D lattices is not so straightforward in the case of large organic molecules. Various laboratories have developed different systems to deal with the BZ of various space groups and we refer the interested reader to individual cases presented in the literature. In our group we have fully treated the case of diamond-like lattices and of hexagonal ice.

6. CALCULATIONS ON 1-D LATTICES WITH DEFECTS

Most of the real organic polymers contain various kinds of chemical, stereochemical and conformational disorder. The concentration and distribution of these defects varies from sample to sample. Since the physical properties of these materials do depend mostly on the type and concentration of defects, lattice dynamical calculations on disordered systems must be carried out as a help in the interpretation of the optical vibrational infrared and Raman spectra.

The basic fact is that, when defects are introduced, the periodicity of the lattice is destroyed and all selection rules are removed and everything in principle can become active. The density of vibrational states becomes an important ingredient in these kinds of works, since the optical spectra represent the dipole weighted (to account for intensities) mapping of the vibrational density of states.

Calculations must then be performed on extremely large systems allowing quick and easy changes of types, concentration and distribution of defects. While the problem is still extremely complex for the case of 3-D crystals, it has been nicely solved when 1-D lattices must be studied (see refs. 5 and 6).

In solid state physics very simple 1-D lattices have been treated mostly using an approach through Green's functions. This theoretical approach is unfeasible in the case of large organic polymers with many atoms per chemical repeat unit. Our group in Milano has applied the so called Negative Eigenvalue Theorem (NET) which can be easily applied to the case of structurally disordered organic polymers. Indeed NET works well for the case of band-matrices; and this is the case of 1-D crystals. Contrary to what happens when Green's functions are used, NET allows to treat molecules without introducing any chemical or structural simplification, thus without loosing the flavour of chemistry so important in organic polymers [5,6].

NET allows to plot the histogram of the density of vibrational states at any desired "numerical resolution". By narrowing the steps of the histogram one can calculate exactly each individual eigenvalue. Eigenvectors can be calculated with the Wilkinson's Inverse Method.

Chain molecules with thousands of atoms with any kind of defect can and have been treated with very reasonable computing time. Our group in Milano has developed computing programs which automatically construct the dynamical matrix with any desired type, distribution and concentration of defects. The only important ingredient to be decided upon is, obviously, the force field.

7. FREQUENCY AND INTENSITY SPECTROSCOPY

It is customary, for those who use the vibrational spectrum, to think only in terms of vibrational frequencies, thus neglecting that vibrational intensities are usually recorded by all spectrometers. This observable has been generally neglected also by specialists in spectroscopy, because of the lack of any usable theoretical model for the understansing of the observed band intensities.

The problem was first tackled in the Sixties at Minnesota (USA) by the school of Professor Crawford, and more recently by Professor Gribov in Russia. In Milano we have reformulated and developed the theoretical model of Gribov, and at present we have provided parameters for the prediction of the vibrational intensities, i.e. the absorption coefficient, of many classes of molecules (for a discussion of the theory see ref. 7; for a list of cases treated see refs. 5 and 7). The theoretical model defines and provides the measured effective equilibrium atomic charges and charge fluxes which occur during the vibration. Many electronic properties of molecules can then be derived. These charges can also be compared (and compare very well) with charges calculated by "ab initio" quantum chemical methods.

Fully automatic computing programs are working in our group in Milano for the prediction of vibrational intensities or for the derivation of electronic molecular parameters. At present infrared spectra (frequencies and intensities) of several classes of molecules can and have been calculated, thus gaining a deeper insight in structural and dynamical problems of small and large molecules.

The effective atomic charges measured from vibrational intensities have been shown to be useful in predicting intermolecular interactions in small and large molecules. These charges can be used when electrostatic pairwise interactions are calculated for the study of structures and stability of molecular systems.

It has been noticed that the calculation of vibrational intensities is an extremely useful way to check the validity of a given molecular force field. Indeed the dipole moment changes during a vibrational motion are weighted by the vibrational amplitudes which in turn are strictly determined by the force field [see Eq.(8)]. It has been shown that well known force fields which predict similar and equally acceptable frequencies for a given molecule do fail in accounting for vibrational intensities.

References

[1] E.B. Wilson, J.C. Decius and P.C. Cross, "Molecular Vibrations", McGraw-Hill, New York (1955).

[2] S. Califano, "Vibrational States", J. Wiley, New York (1976).

[3] G. Zerbi, in "Advances in Applied FTIR Spectroscopy" (M. Mc Kenzie, Ed.), J. Wiley, New York (1988).

[4] G. Zerbi, Applied Spectroscopy Revs. (E.G. Brame, Ed.) 2, 193 (1963).

[5] G. Zerbi, in "Advances in Infrared and Raman Spectroscopy" (R.J.H. Clark and R.H. Hester, Eds.), vol. 11, p. 301, Wiley-Heyden, New York (1984).

[6] G. Zerbi, "Advances in Chemistry" (C.D. Craver, Ed.), American Chemical Society, Washington (1982).

[7] M. Gussoni, "Advances in Infrared and Raman Spectroscopy" (R.J.H. Clark and R.H. Hester, Eds.), vol. 6, p. 61, Heyden, London (1986); M. Gussoni, in "Vibrational Intensities in Infrared and Raman Spectroscopy" (W. Person and G. Zerbi, Eds.), Elsevier, Amsterdam (1984).

Density functional theory and first-principles pseudopotentials: two important tools in solid-state theory

Giovanni B. Bachelet
Dipartimento di Fisica, Universita' di Trento
I-38050 Povo, Trento, Italy

Abstract

The Schrödinger equation for a system of many interacting electrons subject to an external field (typically the electrostatic field of the atomic nuclei) contains in principle all the physics and chemistry of atoms, molecules and solids. Essentially exact solutions of this complicated problem, either based on the Configuration Interaction approach or on the more recent development of Quantum Monte Carlo simulations, are today available (in numeric form) only for light atoms and molecules. An alternative view, very popular in the community of solid-state physicists, is the Density Functional Theory. This theory is well founded and successfully predicts from first principles many properties of real materials. A key ingredient for such a success was the development of new band-structure techniques, especially the introduction of ab-initio pseudopotentials, which optimally replace core electrons in molecular or solid-state calculations. Methods and representative applications will be discussed.

1. Introduction

A rough (and certainly incomplete) picture of first-principles theories in solid-state physics may be given in the form of a flow-chart, shown in Fig.1. Here "first principles" refers to theories based on microscopic laws of interaction which are defined at the start on very general physical grounds. This is opposite to the (often valuable) empirical approach, where the intepretation of a complex system is proposed in terms of a simpler, model system, whose parameters are determined with a best fit to available experimental data.

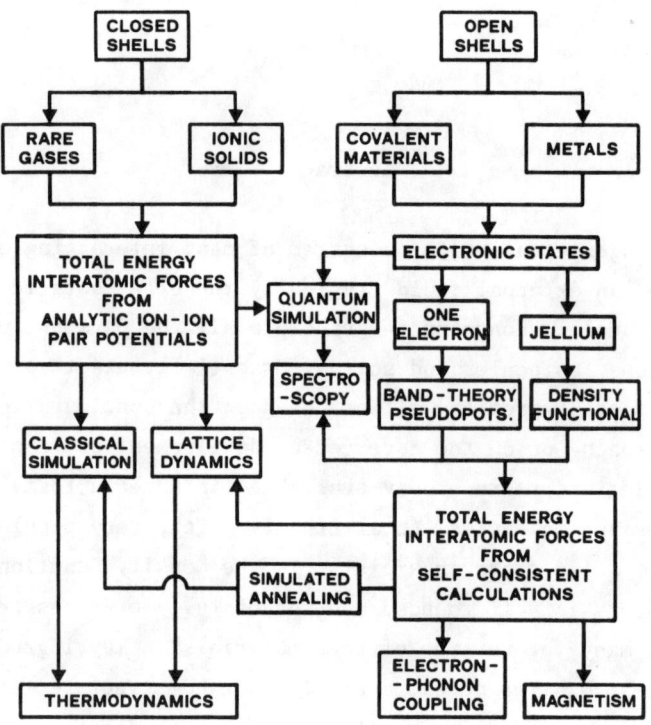

Fig.1 "First-principles" theories in solid-state physics

1.a. Solid-state systems: closed shells and open shells

Let us divide the solid-state systems into two groups; to the left of Fig.1 you can see those which can be built up from closed-shell atoms, like rare-gas atoms; they are especially simple since, for many purposes, their electronic wavefunctions may be forgotten about. Because of the closed-shell configuration their wavefunctions in solids or molecules will not be significantly different from what they are in the isolated atom, and the atom-atom interaction will be satisfactorily described by effective pair potentials without the need of describing the electronic wavefunction explicitly. The same is true for many ionic solids, composed of closed-shell charged ions (like e.g. Na^+ and Cl^- in sodium chloride); in this case the electrostatic interaction, which is absent for rare-gas atoms, is the dominant feature of the effective ion-ion interaction. The ability of evaluating total energies and interatomic forces from the superposition of pairwise interactions (and analytic pair potentials)

[1.1] $$E = \sum_{i<j} v_{ij}(|ri-rj|)$$

represents a major simplification. Analytic expressions for most structural properties of small-size, high-symmetry systems like molecules or bulk solids follow (Bernardes 1958, Tosi 1964). Moreover, numerical studies of the statistical mechanics of large-size, low-symmetry systems (like solid surfaces and interfaces, defects, amorphous materials) become possible. For solids at low temperature the differential properties of the total potential-energy surface near its minimum, explicitly available because of Eq.1.1, contain usually all the relevant information (Lattice Dynamics: see e.g. Maradudin 1966, Jacucci 1986). For solids at higher temperature (and liquids) direct simulations are possible with the technique of Molecular Dynamics, describing the microscopic time evolution of a classical many-body system with given initial conditions (Alder and Wainwright 1960, Rahman 1964, Verlet 1967), or with Monte Carlo methods, based on the idea of sampling configurations from the total potential-energy surface (Metropolis et al. 1953). In all cases one benefits from the validity of Eq.1.1 and the availability of reliable, analytic pair potentials (i.e., no need of including electron coordinates in the description of the atomic interactions).

Unfortunately many molecular and solid-state systems are formed from open-shell atoms (right-hand side of Fig.1). To this class belong both covalent and metallic materials. In both cases it is seldom possible to describe the solid by means of effective potentials. One can of course try to build up effective interatomic potentials whose validity is restricted to a small number of known situations; however, if one seeks predictive power for unknown situations, an explicit description of the electronic wavefunctions is generally needed, because, for the open-shell elements of the Periodic Table, the electronic states are substantially different in isolated atoms, molecules and solids. This poses major conceptual and computational problems, and explains the existence of a wide gap between the stage of development reached by the theory of closed-shell materials (left side of Fig.1) for which the finite-temperature study of a system composed by a hundred independent atoms represents an established routine on today's computers, and the stage of development of the theory of open-shell materials (Fig.1, right side), for which an adequate description of structural properties of e.g. bulk silicon at the absolute zero (just two atoms per unit cell) was greeted as a major breakthrough in 1980 (Yin and Cohen 1980). For the latter group of materials a big theoretical effort was spent since the very early years of quantum mechanics, in two directions: on one hand, the accurate solution of the Schrödinger equation for a single electron in a given external potential (the one-electron theory); on the other, the quantum-mechanical description of a correlated many-electron system (the quantum many-body theory). In the first case one assumes that each electron acts independently of the others and simply knows about their presence in an average way, so that the interaction of an electron with all other electrons and with the nuclei is represented by some effective potential. That of course simplifies life considerably, and one is left with the problem of a single electron in an external field; from this branch come the band theory of crystalline solids, the pseudopotential theory and so on. In the second case one chooses to forget about the presence of ions in a real material and, instead, focuses the attention on fundamental aspects of the electron-electron interaction in a simplified system: a homogeneous electron gas neutralized by a homogeneous background of positive charge (this model is also known as "jellium"; see Fig. 1). From this line of thinking what is most relevant to my presentation is certainly the development of Density Functional Theory (Hohenberg and Kohn 1964, Kohn and Sham 1965), which extended the understanding gained for the homogeneous electron gas to the much more interesting and physically relevant case of an inhomogeneous electron gas. Following the flow-chart in Fig.1 one may notice that the two branches (one-electron theory and the theory of the electron gas) did merge, at some point, to yield the ability of studying from first principles covalent and metallic solid-state systems. The accuracy of these studies, which have boomed in the eighties after the important conceptual development of norm-conserving pseudopotentials (Hamann et al. 1979) and the pioneering results of M.L. Cohen and coworkers (see e.g. the review of M.T. Yin 1985), is such as to predict structural energies, bulk moduli, interatomic forces (thus phonons dispersion curves) and a variety of other physical quantities, the only input being the atomic number of the components and the crystal structure. Initially, this striking success was limited to zero-temperature properties: the motion of ions with simultaneous rearrangement of the wavefunctions appeared as dream; soon after, however, the dream came true, as Car and Parrinello (1985) proposed an elegant and efficient method which unifies Density Functional Theory and Molecular Dynamics (this appears in Fig.1 as "simulated annealing"). Their finding has opened the field of numerical simulation (the microscopic, real-time description of the atomic motion in solids), previously available only for closed-shell systems, to the much larger and more interesting class of covalent and metallic systems. Another box which I want to point out on the flow-chart presented in Fig.1 is that of quantum simulation. This theoretical field of investigation was pioneered by Kalos (see e.g. Kalos et al. 1974) and soon extended to the description of real quantum systems (Anderson 1975; Ceperley and Kalos 1979; Moskowitz et al. 1982; Schmidt and Kalos 1984; Ceperley and Alder 1980, 1984, 1987; Reynolds et al. 1982; see also Binder 1979, 1984 and Kalos 1984). Here no approximation for the electron-electron interaction is adopted, and a direct quantum simulation is carried out by Monte Carlo methods (stochastic sampling of the wavefunction). The state of the art in this field is such that essentially exact results are today available for quantum many-body systems like helium or the electron gas (both homogeneous and inhomogeneous), and even light atoms and molecules. The quantum simulation represents a formidable tool and a promising direction for both solid-state theory and quantum chemistry, but is presently limited by the number of electrons it can treat: suitable extensions of the pseudopotential theory are needed to overcome this limit (Bachelet et al. 1988, Louie 1988, Mitas et al. 1988). As a final comment to the flow-chart (Fig.1) let me emphasize the growing number of links connecting previously separated fields of investigation; this witnesses the excitement, competition and enthusiasm presently experienced by this kind of research.

1.b. One-electron physics

Let me concentrate now on the one-electron physics. The solution of the one-electron Schrödinger equation in a given external potential $V(r)$ (atomic units are used throughout this paper unless otherwise stated)

[1.2]
$$\left[-\frac{1}{2} \nabla^2 + V(r) \right] \psi_i(r) = \varepsilon_i \psi_i(r)$$

is in principle a simple problem, but it may in practice become a rather involved one as soon as one goes from the atom to more and more complicated systems (Fig.2). In the atomic case, due to the spherical symmetry, one is left with a one-dimensional Schrödinger equation for which essentially exact solutions can be obtained by direct numerical integration. For a bulk crystal the problem is reduced to a unit cell (which typically contains a small number of independent atoms) because of translational symmetry; here we already have a quantum-mechanical problem whose complexity is comparable to a molecular problem; it can be turned into a secular problem by the choice of appropriate basis functions. If we finally want to study surfaces, interfaces or impurities, we are left with almost no symmetry: the translational symmetry is lost and even point symmetries can be severely reduced because of lattice relaxation (surfaces, interfaces or impurities may induce substantial

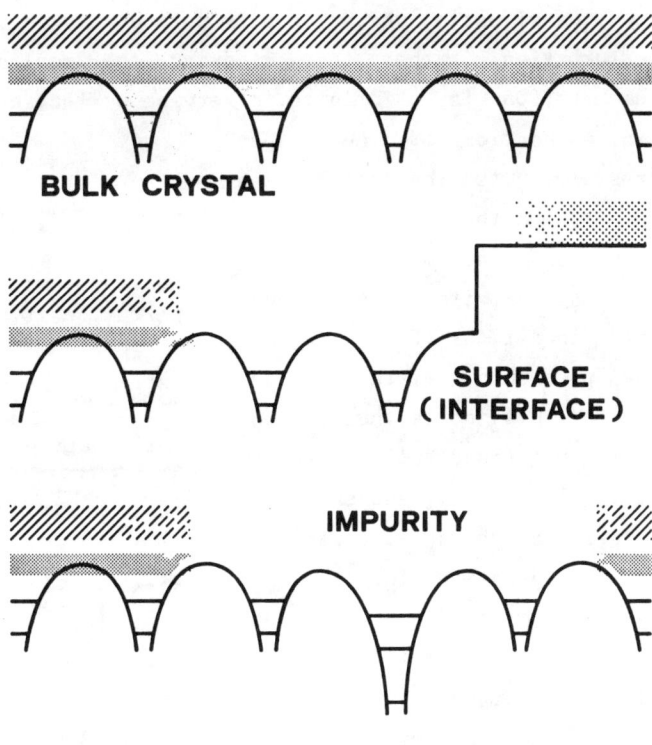

Fig.2 One-electron Schrödinger equation in the atom and in solid-state systems.

atomic displacements in the surrounding bulk cristal); the number of independent atoms is in this case huge, and further theoretical tools are required to bring this kind of problems into some manageable form. One is the Green's function method: one splits the perturbed crystal (Fig.3, above) into an unperturbed bulk part, which still enjoys the translational symmetry, and a perturbation (not necessarily small in amplitude; see Fig.3, middle), and then solves Dyson's equation (see e.g. Callaway 1964). This translates into a practical scheme (i.e., a matrix equation of reasonable size) if the surface, interface or defect perturbation are sufficiently localized in e.g. real space. Another tool which is very popular in computational solid-state theory is the so-called supercell approximation; the trick is here to introduce a new, artificial periodicity so that standard band-structure theory (and computer programs) can be "recycled" to study a situation where the bulk periodicity is lost because of the perturbation. Rather than studying the real system (an isolated perturbation in an infinite, otherwise perfect crystal) one takes a finite portion of the perfect crystal, puts the perturbation in it, and then solves the one-electron problem of such a finite system with periodic boundary conditions (Fig.3, below). The portion of crystal considered for this job is thus the new unit cell (supercell) of an artificial crystal, and should be larger than the spatial range of the perturbation for the new sytem to give a good approximation of the isolated perturbation. The size of the supercell is in practice limited by the available computer power and

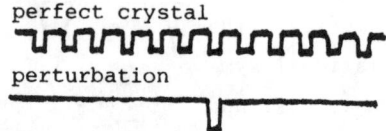

real system:
perturbed crystal

perfect crystal

perturbation

supercell approximation

Fig.3 Green's function vs. supercell (see text)

memory, but the simultaneous development of more efficient theoretical tools and large-scale computers has rapidly pushed up this limit from a few atoms to more than a hundred; further growth is probably just a matter of time. Depending on the nature of the problem under examination these and many other other theoretical tools, devised to yield accurate solutions of the one-electron Schrödinger equation in a solid-state system, have been developed over the years by a large number of active research groups (see e.g. Bachelet 1986, and refs. therein). The effort has been gigantic between the early days of quantum mechanics and today; the state of the art in this field is today such that widely different methods give answers in quantitative agreement with each other for benchmark physical systems (see next sections 3,4). The historical motivation for such an effort is that the one-electron theory, over the past thirty years, has produced key contributions to the understanding of electrical and optical properties of real materials (see e.g. Ashcroft and Mermin 1976).

1.c. Jellium: exchange and correlation energy

As mentioned, exact results are available for the jellium model (homogenous electron gas with a positive neutralizing background). For the exchange part the energy versus density is a known, analytic result: the solutions of the Hartree-Fock equations are plane waves for such a model system (see e.g. Ahscroft and Mermin 1976). A complete description beyond Hartree-Fock (i.e. beyond exact exchange) amounts to determining the "electron-electron correlation energy" (by definition this is the difference between the exact and the Hartree-Fock energy), and has been attempted first by analytical approaches (see e.g. Singwi et al. 1970; Hedin and Lundvist 1971), and, more recently, by direct, numerical quantum simulation of the homogeneous electron gas (Ceperley and Alder 1980). In the latter work the energy of the homogeneous electron gas was obtained for a wide range of densities (see Fig.4); this is probably best available estimate for the exchange plus correlation energy in jellium.

Fig.4 Energy of the homogeneous electron gas as a function of $r_s = (3n/4\pi)^{1/3}$, where n is the density, referred to the lowest boson state (after Ceperley and Alder 1980).

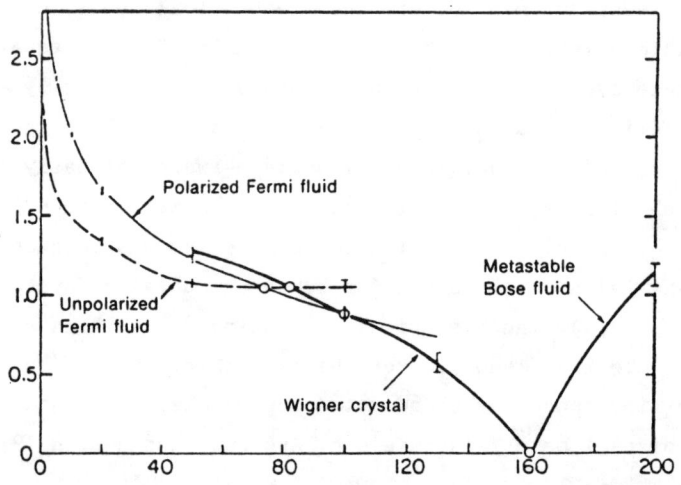

1.d. Jellium, one-electron theory, and real materials

What's the relation of jellium to real materials? One would expect that the jellium model is a very poor approximation of a real material, since it completely neglects the nuclear potentials and replaces them with a shapeless, uniform neutralizing background of positive charge. However the free-electron theory of conduction in metals, which is based on this very model, is already capable of giving very sensible results. Moreover, after adding to the jellium model a suitable, weak periodic potential perturbation with the correct crystal symmetry, one obtains results in qualitative and often quantitative agreement with experiment for a number of physical properties of, say, simple metals like sodium, or even semiconductors like silicon (see e.g. Ziman 1969; Heine et al. 1970; Bassani and Pastori Parravicini 1974). In both cases the theory reduces the complex quantum-mechanical problem of interacting electrons in a real material to one-electron equations of the type of Eq.1.2. Since the atoms composing these materials have a lot of core states, and since the true electronic density is far from being constant (in fact, it undergoes enormous spatial variations in space from the inner core region to the valence region, and, in covalent systems, from the bond region to the interstitial region), two important and related questions arise. The first concerns core electrons. Why is it possible to describe some properties like conduction or cohesion in simple metals and predict electrical and optical properties of semiconductors without any reference to the inner core? The second concerns the electron-electron interaction. Why can we understand many materials in terms of the simple one-electron physics underlying Eq.1.2, in spite of significant electron-electron interactions?

2. Density Functional Theory (DFT)

The success of the one-electron description of solids rests on firm theoretical grounds, and does not result from an accidental, fortunate coincidence. There are many ways to argue that Eq.1.2 must be a good approximation of the complete many-body problem, but a general and rigorous argument can be found in two famous papers on the inhomogeneous electron gas (Hohenberg and Kohn 1964, Kohn and Sham 1965; hereafter refereed to as HKS). HKS demonstrated that (a) the ground-state energy of an interacting many-electron system subject to an external potential $V_{ext}(r)$ is a <u>unique functional</u> of the (spatially varying) electronic density $n(r)$, and (b) minimization of the energy functional with respect to the density results in a set of one-electron equations like Eq.1.2. Although this finding may sound, at first, like a straightforward extension of the homogeneous case, where all physical quantities are <u>functions</u> of the (spatially constant) electronic density, it is far from being a trivial result, and its consequences are of great importance. That the electronic density must be some functional of the external potential is easy to prove and accept; that the converse is also true, namely that the external potential (within an additive constant) is a unique functional of the density, is certainly less obvious. The latter statement is precisely the key theorem of density functional theory; its proof is based on a "reductio per absurdum" (Hohenberg and Kohn 1964). In spite of the numerous attempts to disprove it, the theorem has successfully resisted over twenty years of theoretical debate, during which it was also extended from the simple, original case of a nondegenerate ground state at zero temperature, to more general situations (Mermin 1965, Gunnarsson and Lundqvist 1976). Let us briefly recall some results of the HKS theory which are relevant to this presentation. Given a certain external potential $V_{ext}(r)$, the ground-state electronic energy is a unique functional of the electronic density: $E = E_{V_{ext}}[n]$. From the ground-state electronic energy we subtract the integral of the external potential times the density and obtain a new, <u>universal</u> functional of the electronic density (the part explicitly depending on V_{ext} has been subtracted out):

$$[2.1] \qquad G[n] = E_{V_{ext}}[n] - \int d^3r \; n(r) \; V_{ext}(r)$$

We now subtract the Hartree (electrostatic) energy and the kinetic energy of the (noninteracting) electron gas corresponding to that density. Both of these terms are universal density functionals; for the former this property is self-evident; for the latter, which we call $T_s[n]$, it was readily demonstrated by HKS. After this subtraction we are left with another universal density functional

$$[2.2] \qquad E_{xc}[n] = G[n] - \int d^3r \; d^3r' \; \frac{n(r) \; n(r')}{|r-r'|} - T_s[n]$$

For reasons which will be evident in a moment, this functional is called "exchange-correlation energy".

2.a. Kohn-Sham (KS) equations

To minimize with respect to the density the total electronic energy

[2.3]
$$E_{V_{ext}}[n] = T[n] + \int d^3r \, V_{ext}(r) \, n(r) + \int d^3r \, d^3r' \, \frac{n(r) \, n(r')}{|r-r'|} + E_{xc}[n]$$

[the number of particles $N = \int n(r) d^3r$ being fixed] corresponds to solving a single-particle Schrödinger equation (Eq.1.2) where the potential $V(r)$ is given by the condition of self-consistency:

[2.4]
$$n(r) = \sum_{i=1}^{N} |\psi_i(r)|^2$$

[2.5]
$$V(r) = V_{ext}(r) + \int d^3r' \, \frac{n(r')}{|r-r'|} + \frac{\delta E_{xc}}{\delta n(r)}$$

The sum in Eq. 2.4 runs over the occupied single-particle states. Eq. 1.2, together with Eqs. 2.4, 2.5, are known as Kohn-Sham (KS) equations. They reduce the many-body problem to a one-electron problem: besides the electrostatic potential generated by the electronic charge distribution, all the quantum effects of the electron-electron interaction are now lumped into the exchange-correlation potential $V_{xc}(r) = \delta E_{xc}/\delta n(r)$, the functional derivative of the exchange-correlation energy with respect to the density (last term in Eq. 2.5). We observe that (a) so far <u>no approximation</u> has been made, so that this theory is in principle capable of dealing with electron-electron interactions (electrostatic, exchange and correlation) exactly, (b) unlike the method based on Configuration Interactions the theory can be expressed in a <u>simple one-electron form</u> very similar to the Hartree-Fock equations (this is in fact the origin of the name "exchange-correlation": that part replaces the exchange energy in the Hartree-Fock equations, but also contains all the correlations), and finally (c) unlike Hartree-Fock, the exchange-correlation potential is local: it acts in a simple multiplicative fashion on the wavefunction. The last point is not a minor one, and implies that the computational time required to solve the Schrödinger equation with a basis set of N functions scales like N^3 for the KS equations (the dependence is N^4 for Hartree-Fock). As mentioned, the theory in its most general form applies to a variety of situations not included in the original work of HKS (see Mermin 1965, Gunnarson et al. 1976,1979, Hybertsen and Louie 1985, Godby et al. 1986, Oliveira et al. 1988). Of course nothing is perfect. The use of Density Functional Theory in practical applications (like the calculation of physical properties of atoms, molecules and solids) requires the knowledge of the exchange-correlation functional. The HKS theory demonstrates the universality of such a functional, but unfortunately does not provide its form.

2.b. Local Density Approximation (LDA)

Notwithstanding the theoretical effort spent since 1966, a general, explicit form of the exchange-correlation functional has not been found. This is where approximations come into play: the best thing which could be done, up to now, was to conceive (and use in practical calculations) approximate expressions for special classes of density distributions. The simplest one is certainly the Local Density Approximation, which applies, as already shown in the original HKS papers, to density distributions whose spatial variations are either slow or small in amplitude. This amounts to writing the exchange-correlation functional in the form

$$[2.6] \qquad E_{xc}[n] = \int d^3r \; n(r) \; \varepsilon_{xc}^{hom}(n(r))$$

namely, to assume that locally the inhomogeneous electron gas behaves like a homogeneous gas; in other words the exchange-correlation energy density ε_{xc} at point r exclusively depends on the electron density at point r, and the functional dependence is taken from the homogeneous electron gas:

$$[2.7] \quad \varepsilon_{xc}^{hom}(n) = \varepsilon_x^{hom}(n) + \varepsilon_c^{hom}(n) = -\frac{3}{2}\left(\frac{3}{8\pi}\right)^{1/3} n^{1/3} + \varepsilon_c^{hom}(n)$$

where the exchange part ε_x corresponds to the Hartree-Fock result, which is analytically known in the homogeneous case, and the correlation part ε_c can be obtained from state-of-art results for the homogeneous electron gas (Ceperley and Alder 1980, Perdew and Zunger 1981). This approximation has been widely used in atomic, molecular and solid-state applications by physicists. But in fact, does (and should) it work for real systems, where the spatial variations of the electronic density are neither slow nor small in amplitude? Many theoretical chemists appear somewhat skeptical about local-density functionals; in Sect.4 some answers will be suggested.

2.c. Self-consistent loop and simulated annealing

Let us see first how to actually find the ground-state energy and density of a real system, composed of nuclei and interacting electrons, according to DFT. The prevailing strategy was until recently to solve the KS equations by iterating to self-consistency a two-step process: in the first step one solves the one-electron Schrödinger equation (Eq.1.2) for a given "input" potential; in the second step one builds up the charge density from the wavefunctions of the occupied one-electron states (Eq.2.4), and from it the Hartree (electrostatic) and the exchange-correlation potential (Eqs. 2.5-2.7). Then, in a feedback process, the resulting "output" potential (Eq.2.5) becomes the "input" potential for the next iteration (actually some attenuated feedback is needed in the beginning, to damp charge fluctuations). Typically in the first step one uses a basis set (either localized functions like gaussians, Slater orbitals, muffin-tin orbitals, or extended functions like plane waves, or plane waves augmented with localized functions etc.) to solve the differential Schrödinger equation, and ends up in a NxN secular problem which generally implies a computational effort of order N^3. Here N is the number of basis functions, which is proportional to the number of independent atoms N_a of the system under examination. Since a certain variational flexibility is required for accurate calculations, the number of basis functions per atom cannot be smaller than, say, ten or twenty, and the N^3 dependence of computer time represents a serious limitation for the study of large molecules, or solids with a large unit cell.

Alternative ways to find the minimum of the HKS energy functional without having to diagonalize a matrix were also explored, and found successful applications in the second half of the eighties. One of them is intrinsic to the new unified approach to DFT and Molecular Dynamics (Car and Parrinello 1985); the idea is to associate to the electronic degrees of freedom some fictitious dynamical variables, and then perform a simulated annealing (Cerni 1982, Kirkpatrick et al. 1983). For a system of N_a independent atoms the computational effort required to find the ground-state density and total energy depends in this case on terms proportional to M^2N, MN^2 and $MN\ln N$, instead of the N^3 dependence implied by standard diagonalization routines (N is the number of basis functions, while M is the number of occupied states). Both N and M are proportional to the number of independent atoms N_a (thus also in the Car-Parrinello approach the third-power dependence on N_a will represent the ultimate limiting factor); the point is that normally M<<N, so that, given a certain computer power, the size of systems which can be studied with this method is much larger than with standard diagonalization techniques (about 150 atoms with computers of the Cray-XMP family). Of course the great appeal of the Car-Parrinello method is that, besides this computational advantage, it allows to study the atomic motion (Car and Parrinello 1985, 1988). If the atoms are not allowed to move, then other minimization techniques which are even more economic are applicable to the electronic-structure problem (Stich et al. 1988; practical aspects of the use of nonlocal pseudopotentials are also discussed there).

3. Pseudopotential Theory

Core electrons, very localized near the nucleus, seldom take part in the binding process of atoms in molecules and solids. The nature of the chemical bond is mostly determined by valence electrons, as witnessed by the periodic properties of elements in the Mendeleev Table. Thus the assumption that core electrons remain practically unchanged in molecules and solids ("frozen-core" approximation) is usually a good one. However the details of valence orbitals are such that two elements of the same group of the Periodic Table show similar, but not identical behavior; these details, needed for a quantitative understanding of the chemical bond, are crucially affected by the nature of the underlying core orbitals. The requirement of orthogonality to core states pushes valence wavefunctions away from the core region; core electrons do not directly participate in the binding process, yet valence electrons "know" about their presence because of mutual orthogonality. Is it possible to get rid of the core electrons and represent their influence on valence electrons in terms of some effective - or "pseudo" - potential? Pseudopotential theory dates back to the late fifties (Antoncik 1959, Phillips and Kleinman 1959) when it was realized that the orthogonality to core states is equivalent to a fictitious repulsive potential acting on valence wavefunctions only; this repulsive part, localized in the core region, nearly cancels the attractive coulomb potential of the atomic nucleus; in general, then, one can remove core electrons along with the coulomb singularity of the nuclear potential, ending up with a weak pseudopotential which acts on valence electrons only. That's why, to answer two questions raised in Sect. 1, many physical and chemical properties of atoms in condensed matter can be well described without any apparent reference to core electrons, and a weak perturbation of the electron gas may describe to a good approximation some classes of real materials. We are now left with the problem of how to actually replace a full-core atom (the real system) with a pseudoatom where only valence electrons are present, and with the more fundamental question whether such a substitution can be at all based on first principles (i.e. on some exact relation to the full--core atom), or inevitably leads to a rough, empirical model which cannot be really trusted as a reliable valence atom in the context of self-consistent molecular or solid-state calculations. To clarify this point a concise history of pseudopotentials may be of help.

3.a. Empirical pseudopotentials

In the sixties and early seventies many important properties of electrons in crystals were understood with an empirical use of the concept of pseudopotential (see Heine et al. 1970). A weak periodic pseudopotential, representing the crystal potential felt by valence electrons in, say, a metal or a semiconductor, was represented in terms of a few Fourier components (only a few plane waves were thus needed to solve the corresponding Schrödinger equation). Rather than bearing fundamental relations with the full-core atoms which form the crystal under consideration, the Fourier components of the pseudopotential were used as disposable parameters to fit or interpolate some measured crystal property (optical spectra, transport properties, cohesive energies etc.). In this way it was for example demonstrated the great importance of crystal symmetry on the optical properties of solids. Applications were limited to simple metals and semiconductors, and it was recognized that first-row elements like carbon or oxygen and elements with valence d-orbitals like transition elements or noble metals are not well described by a weak, local pseudopotential (Bassani and Pastori Parravicini 1974). Even to produce a perfect fit of all the measured optical properties of semiconductors it was suggested that a nonlocal (angular-momentum dependent) pseudopotential may be required; this result was part of an extensive work of Chelikowsky and Cohen (1976), which probably represents the highest point reached by this type of approach.

3.b. Model pseudopotentials

In the seventies an important step in pseudopotential theory was represented by model pseudopotentials constructed and used in the framework of the local-density approximation. Instead of using pseudopotential parameters to directly fit the measured solid-state properties the idea was to fit atomic properties and then use the obtained ionic potential to predict solid-state properties. A smooth parametrized bare-ion pseudopotential with a $-Z_v/r$ tail was screened with Z_v valence electrons to yield e.g. a neutral ground-state pseudoatom; the pseudopotential parameters were then varied until the self-consistent atomic eigenvalues agreed with the (calculated or measured) valence energies of the corresponding true atom. The bare-ion pseudopotential was then plugged into a solid-state environment: bulk crystal, surfaces, lattice vacancies etc. Here the self-consistent rearrangement of valence electrons was studied. This effort was especially valuable in the interpretation of optical spectra of low-symmetry systems (see e.g. Appelbaum and Hamann 1978, Cohen 1979); at the same time experience taught that the approach wasn't accurate enough to predict structural energies. In the case of silicon, for example, the experimental phonon dispersion curves could be reproduced only if empirical spring constants were added to the force constants as obtained from self-consistent electronic structure calculations (Wendel and Martin 1979); moreover, in contrast with nature, calculated total energies for different crystal structures would not energetically favor the diamond structure (Ihm and Cohen 1979). Of course cohesive energies, structural energy differences and phonon energies are tiny fractions of the total (valence) energy in the solid, so that a great accuracy is required for this kind of prediction. But it was soon recognized that, besides numerical inaccuracies, the most important source of such a failure was the lack of explicit relations of the pseudowavefunctions with the full-core atom: the model pseudopotentials could reproduce the atomic valence energies but not the tails of the valence wavefunctions. This was the motivation to attempt the construction of ionic pseudopotentials from first-principles, namely, with explicit relations with the full-core atoms both for energies and wavefunctions (Shaw and Harrison 1967, Kahn and Goddard 1972, Melius and Goddard 1974, Topp and Hopfield 1974).

3.c. First-principles pseudopotentials

Let us see which desirable properties should characterize a first-principles pseudoatom. In the full-core atom, the potential diverges in the origin and the valence wavefunctions have nodes, because of orthogonality to inner core states; in a pseudoatom, core states are eliminated and only valence wavefunctions with no nodes are present. For a given valence configuration often called reference atomic configuration or reference state (typically the ground state of the neutral atom), the pseudopotential should reproduce exactly the valence eigenvalues and eigenfunctions of the original full-core atom (the latter only outside a certain "core radius" r_c placed between the outermost node and the outermost maximum of the full-core valence wavefunction). Moreover, for the pseudopotential to be useful in situations other than the reference atomic state for which it was constructed, the pseudoatom should be transferable: it should behave like the original full-core atom in a variety of different chemical situations, and the response of its pseudocharge density to a molecular or solid-state environment should be as close as possible to the response of the valence charge density of the original full-core atom. Because of well-known relations connecting logarithmic derivatives, scattering amplitudes and matching conditions (see e.g. Landau and Lifschitz 1966) the same two requirements (exact reference state and optimum transferability) can be rephrased in terms of logarithmic derivative versus energy, the logarithmic derivative of the wavefunctions being taken at some radius outside the core, in the region of space which is relevant for the formation of the chemical bond ($r > r_c$). There

(i) we want the pseudo and full-core potentials to have the same logarithmic derivative at a certain reference energy (for example the bound-state energy in the reference atomic configuration) for each valence state; and (ii) we also require that the pseudo and full-core logarithmic derivatives follow each other as closely as possible away from this energy. The energy range over which the logarithmic derivatives are identical (or at lest very close) is ideally 15-30 eV: in typical molecular or solid state situations bound-state energies are shifted away or spread around the atomic eigenvalue by a similar amount, due to bonding and banding effects. In the late seventies (Joannopoulos et al. 1977, Topiol and Zunger 1977, Redondo et al. 1977, Harris and Jones 1978, Zunger and Cohen 1978, Christiansen et al. 1979, Starkloff and Joannopoulos 1979) a growing consideration was devoted to the tails of the wavefunctions in the pseudoatom compared to the tail of the corresponding valence wavefunctions in the full-core atom. The tails of the valence wavefunctions are very important for the chemistry of atoms. It became clear that the reproduction of the valence eigenvalues for a given reference configuration of the full-core atom was not sufficient to reproduce the wavefunctions outside the core. From the form of the Schrödinger equation one immediately sees that the identity of the eigenvalue only implies that pseudo and full-core wavefunctions be proportional outside the core region ($r>r_c$). If the empirical pseudopotentials of the sixties, directly constructed to fit some measured property in the solid, did not bear

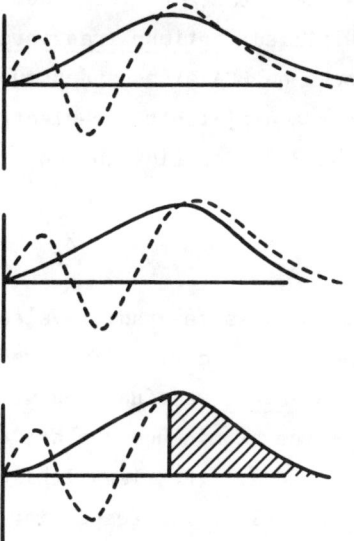

Fig.5 Pseudowavefunctions from empirical (top), model (middle) and norm-conserving (bottom) pseudopotentials (see text).

any explicit relation with the full-core atom (top of Fig.5), the model pseudopotentials of the seventies (middle panel of fig.5) were such that their pseudowavefunctions, even for the reference atomic configuration, were wrong by a factor of proportionality outside the core region ($r > r_c$). The same problem was inevitably encountered by "OPW-like" pseudopotentials (see e.g. Heine et al. 1970). The main goal in the late seventies was therefore to bring this factor of proportionality very close to one, and thus obtain, outside the core, pseudowavefunctions as close as possible to the original valence wavefunctions of the full-core atom (at least for the reference atomic configuration). In this context Hamann, Schlüter and Chiang (HSC) showed (1979) that it is possible to construct smooth pseudopotentials that <u>exactly</u> reproduce the tail of the valence wavefunctions for a reference atomic configuration, and - more important - they demonstrated that, once this requirement is fulfilled, <u>optimum transferability</u> follows. [In what follows I will focus on LDA atoms and pseudoatoms; similar ideas apply to Hartree-Fock atoms] Their one-electron bound-state wavefunctions (Eqs. 1.2, 2.4, 2.5), outside the core (for any $R > r_c$), satisfy the identity

$$[3.1] \qquad \chi_L^{PS}(E=E_L,R) \equiv \chi_L^{FC}(E=E_L,R)$$

$r\chi_L(E,r)$ is the radial valence wavefunction for the pseudo (PS) and the full-core atom (FC), respectively; at $E=E_L$ it gives the valence <u>bound-state</u> wavefunction with eigenvalue E_L. Eq.3.1 implies that, at the atomic bound-state energy E_L, for any radius $R > r_c$ outside the core, the logarithmic derivative of the wavefunction is also <u>identical</u> for corresponding full-core and pseudo valence states

$$[3.2] \qquad \left(\frac{\chi'_L}{\chi_L}\right)^{PS}_{E=E_L,R} \equiv \left(\frac{\chi'_L}{\chi_L}\right)^{FC}_{E=E_L,R}$$

Away from E_L discrepancies are only second order in the energy $E-E_L$. This very desirable property comes about because the integral

of the wavefunction squared is related to the energy derivative of
the logarithmic derivative of the wavefunction

$$[3.3] \quad \left.\frac{d}{dE}\left(\frac{\chi'_L(E,R)}{\chi_L(E,R)}\right)\right|_{E=E_L} \equiv 2\int_0^R dr \left(\frac{\chi_L(E=E_L,r)}{\chi_L(E=E_L,R)}\right)^2$$

If both the FC and the PS wavefunctions satisfying Eq.3.1 are correctly normalized, then, by Eq.3.3, the first energy derivative of the logarithmic derivative will also be identical at the bound-state eigenvalue

$$[3.4] \quad \left.\frac{d}{dE}\left(\frac{\chi'_L(E,R)}{\chi_L(E,R)}\right)^{PS}\right|_{E=E_L} \equiv \left.\frac{d}{dE}\left(\frac{\chi'_L(E,R)}{\chi_L(E,R)}\right)^{FC}\right|_{E=E_L}$$

The price to be paid for fulfilling Eqs. 3.1-3.4 is that the HSC pseudopotentials are L-dependent, i.e. nonlocal, and their use in molecular or solid-state calculations requires the use of angular projection operators. The L-dependence reflects the physical fact that different partial waves feel the effect of orthogonality to different core states. The exact fulfillment of the above conditions is relevant only for angular momenta which correspond either to occupied valence orbitals in the atom, or to empty orbitals whose energy is low enough to be potentially involved in the bonding process. For most atoms, with the exception of rare earths and lanthanides, excellent norm-conserving pseudopotentials are obtained by imposing Eq.3.1 to s,p and d pseudowavefunctions; for many atoms it may be sufficient to fix only s and p waves (see Bachelet et al. 1982). Norm-conserving pseudopotentials, which exactly reproduce, for the reference atomic state, the "orthogonality hole", pushing away from the core region as much valence charge as the effect of orthogonality to inner core states would push in the original full-core atom (bottom panel of Fig.5), seem thus to enjoy most desirable properties of the ideal pseudopotential.

3.d. Examples and tests of transferability

Rather than showing how to construct them, which can be found in the original papers (the method has also been extended to deal with heavy atoms where the Dirac equation applies: see Kleinman 1980, Bachelet and Schlüter 1982; norm-conserving pseudopotentials have been then tabulated for the whole Periodic Table by Bachelet, Hamann and Schlüter in 1982), let us look at some representative cases which demonstrate their properties, especially their high transferability and their ability to faithfully reproduce the corresponding full-core atoms in a variety of chemical situations.

In Fig.6 we see the ion-core pseudopotential for Au. The 5d electrons experience a strong attractive potential. This potential, however, is considerably less attractive than for 3d or 4d transition or noble elements due to increased orthogonality effects. Strong spin-orbit splitting effects are present, with decreasing amplitude for p, d and f electrons. The Au

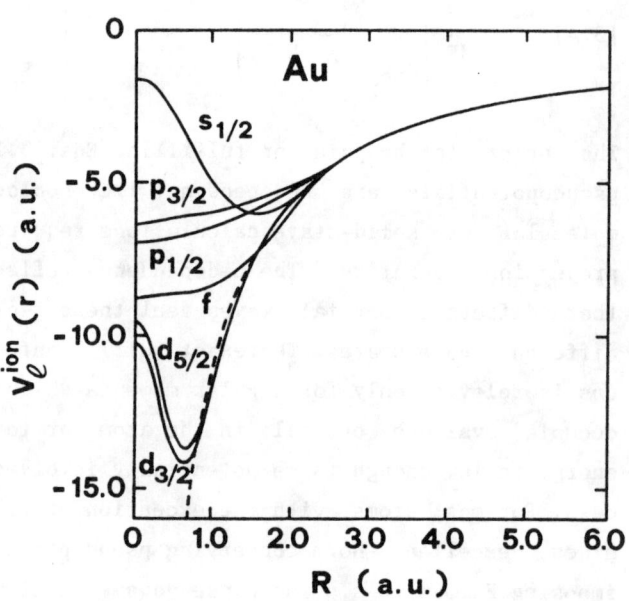

Fig.6 Relativistic norm-conserving pseudopotential for Au. The pseudopotential is nonlocal and depends on L and j. The dashed line is the coulombic tail.

valence wavefunctions are illustrated in Fig. 7. The 6s and 5d wavefunctions are calculated from the atomic ground state ($5d^{10}6s^1$) while the 6p and 5f states are obtained from excited states, re-

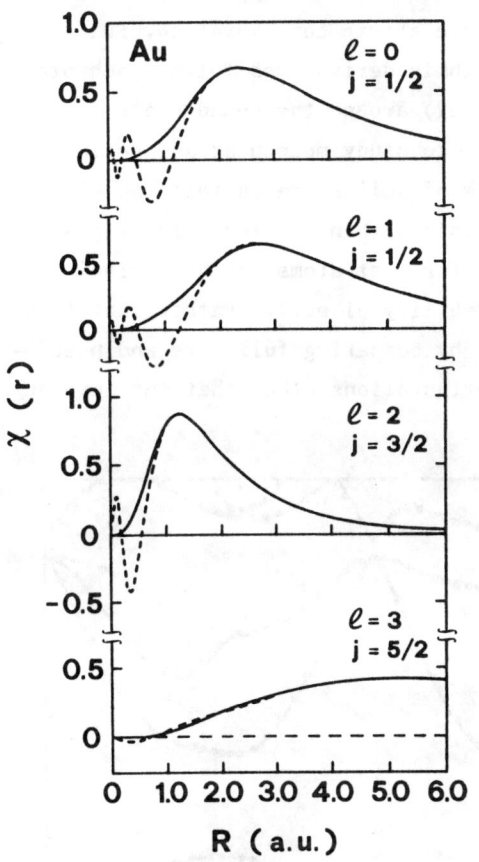

Fig.7 Comparison of full-core and valence wavefunctions of Au. Dash:full-core atom; solid: pseudoatom. $\chi(r)$ is $r\psi(r)$ where ψ is the radial wavefunction. Of each relativistic doublet only the component with highest j-value is shown. Atomic units are used.

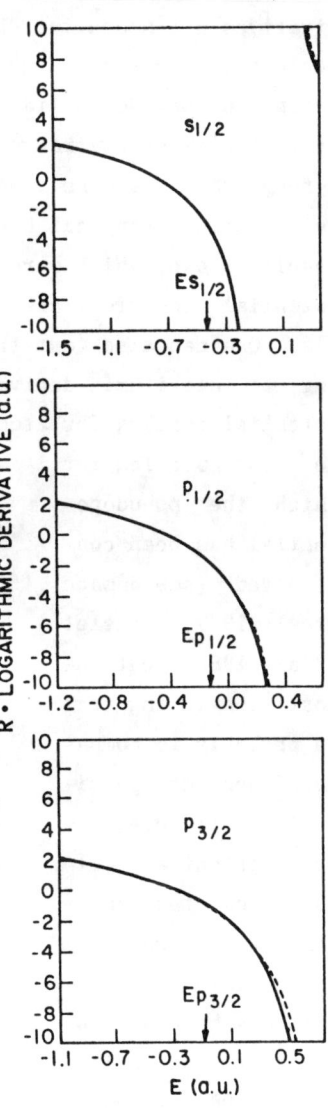

Fig.8 Logarithmic derivatives for Pb (see text)

spectively. Only the j=L+1/2 spin-orbit components are shown. The multishell behavior, typical for transition, noble, and rare-earth elements is clearly visible. In Fig. 8 the s and p logarithmic derivatives are shown for Pb as a function of energy. Arrows show the bound-state energy in the reference atomic configuration. Full-core (dash) and pseudo (solid) logarithmic derivatives follow each other over a wide energy range (20-30 eV) around the bound-state energy, a range which is more than enough to study molecular and solid-state situations (typical bandwidths of solids are in this range). The results shown, which were unfeasible within previous, local pseudopotential theories, are typical for most atoms in the Periodic Table. One can also test transferability directly, rather that looking at logarithmic derivatives, by comparing full-core and pseudopotential results for atomic configurations other that the reference configuration for which the pseudopotential has been constructed (see Hamann et al. 1979, Bachelet et al. 1982), but the most convincing test is probably to compare (whenever possible) full-core and pseudopotential calculations for an identical, benchmark solid-state system. This has been done by Bachelet and Christensen (1985) for GaAs (in Fig.9 the band structure is shown)

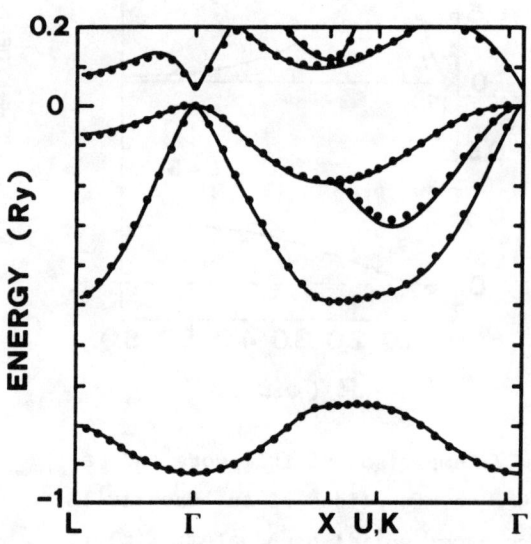

Fig.9 Band structure of GaAs. Solid: full-core (LMTO method). Dots: norm-conserving pseudopotentials. Both are based on LDA.

and by Hamann (1979) for silicon (in Fig. 10 I show his results for the valence charge density along the <110> plane of crystalline silicon). The results shown in Figs. 9 and 10 were all based on the local-density approximation (LDA), described previously. It is evident from Figs. 9, 10 that, when comparing the full-core atom with the pseudoatom, both valence eigenvalues and valence charge distributions are in perfect agreement. These tests are representative of a much larger experience on norm-conserving pseudopotentials, and give pictorial evidence of their reliability and chemical transferability. It is also interesting to observe that, when constructing norm-conserving pseudopotentials in the Hartree-Fock approximation instead of LDA, one ends up, after "unscreening" the valence charge, with essentially identical ionic pseudopotentials (Gygi and Baldereschi 1986).

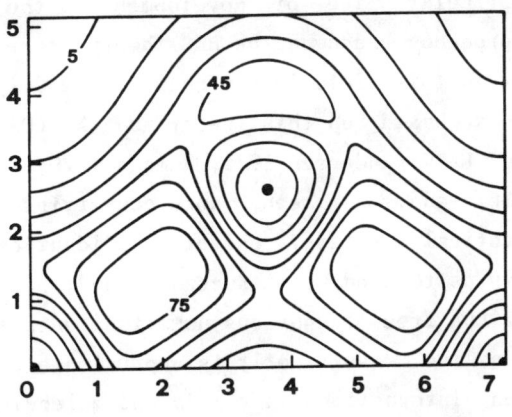
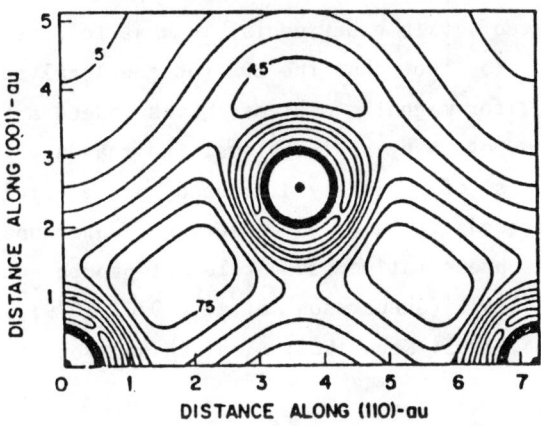

Fig.10 Charge density along the <110> plane for Si. Top: pseudo. Bottom: full-core.

4. Representative results

The pseudopotential method just discussed represents only a particular line of development in the context of first-principles approaches based on the HKS theory. More generally, the state of the art in this field is now such that widely different numerical approaches based on this theory and the LDA (for example full-core LMTO and LAPW: Andersen 1975, Hamann 1979; full-core LCGO, Harmon et al. 1982; and, as seen, norm-conserving pseudopotentials) yield, for identical systems, essentially identical results. This finally allows us to study both successes and, possibly, intrinsic failures of the LDA free of spurious numerical uncertainty. We only need to compare the results of this approximate description of electron-electron interactions in solids and molecules with the results of exact theories (seldom available, especially for solids) or, in the lack thereof, with experiments. In what follows a few examples are intended to show that the LDA (or the local-spin-density approximation, LSD, for magnetic systems) gives indeed a rather successful description of many real systems; the reason why it works well beyond the limits of validity discussed in the original HKS papers is now understood to a good extent, but unfortunately exceeds the space of this presentation; some relevant papers can be found in the reference list (Gunnarsson et al. 1976, 1979; Gunnarsson and Jones 1980, 1984, 1985; Kohn 1983; Almbladh and von Barth 1985).

4.a. The Ozone molecule

In a number of full-core LSD calculations Jones (1984, 1985, 1986) has recently shown both the power and the limits of such an approach for small molecules. A striking case are his results for the triatomic ozone molecule (1984, Fig. 11). Here the energy surfaces of the molecule are shown as a function of the bond angle at the equilibrium bond length, for states of different symmetry. As far as <u>relative energies</u>, <u>interatomic forces</u>, and <u>force constants</u>

are concerned, the results are in remarkable agreement with CI calculations (whenever available), and the large set of new results sheds new light on the physics of such an important molecule; however the absolute binding energy is overestimated by approximately 25% with respect to both experiment and CI calculations; this, together with other results, suggests that LDA and LSD have a systematic tendency to overestimate absolute binding and cohesive energies in s-p bonded molecules and solids.

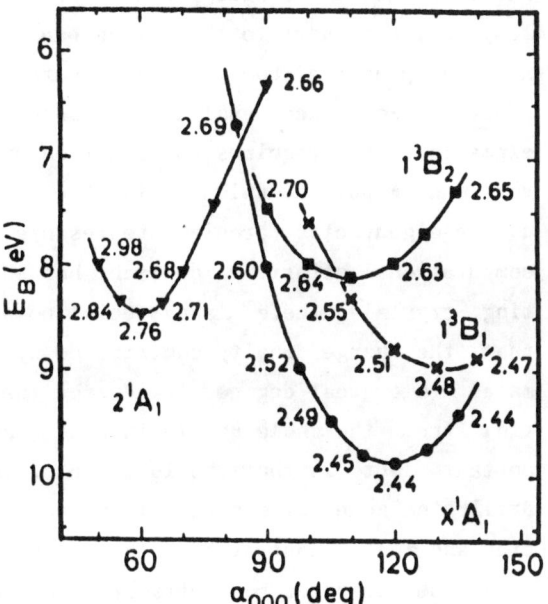

Fig.11 Energy vs. bond angle for the triatomic ozone molecule. The various curves correspond to different symmetries and the equilibrium bond length is marked for each angle (Jones 1984)

4.b. Bulk silicon

As mentioned in Sect.2, the standard use of LDA consists in the solution of the Kohn-Sham equations under the requirement of self-consistency between the charge density corresponding to the occupied states and the electrostatic and exchange-correlation potential. There is no doubt that the availability of large computers of increasing memory and speed has been an essential ingredient for these calculations. But this would not be sufficient without the parallel development of the theory. Besides the pseudopotential method, many other theoretical progresses were crucial to the feasibility and success of self-consistent calculations of electronic states in a crystal; they cannot be even listed here. As an example of this vast theoretical work I'll pick an important theorem due to Baldereschi (1973); its results and extensions are today used in most self-consistent crystal calculations for insulators. In a crystal the evaluation of the charge density requires in principle the knowledge of the single-particle wavefunctions of all occupied states, which amounts to the filled bands over the whole Brillouin zone. In practice one would divide the Brillouin zone into small k-space regions and evaluate the integral over k in terms of finite elements. This requires the solution of the Kohn-Sham equations for very many k-points, which is inconvenient since in general the eigenvalue-eigenvector problem represents the largest portion of the computational effort. Baldereschi has shown in 1973 that for insulating crystals there is a special point in the Brillouin zone such that the charge density contributed by this single k-point approximates to a great degree of accuracy the total charge density resulting from the whole Brillouin zone (the mean-value point). The important fact is that the location of this special point within the Brillouin zone is exclusively dictated by the symmetry of the lattice and independent of any other crystal property. Due to this finding the computer cost of self-consistent calculations for insulators and semiconductors, which represent of course an important and active field of investigation for materials science, was cut down by one to two orders of magnitude. Bulk silicon, in particular, has always been one of the most popular benchmark systems for the theory of electron states in solids because of obvious connections between fundamental advances in the understanding of its properties and the development of semiconductor technology. The first great success of the combined use of norm-conserving pseudopotentials and the local--density approximation was indeed obtained for bulk silicon by Ming Tang Yin and Marvin L. Cohen (1980). With this test case they demonstrated both the feasibility and the reliability of these two theoretical tools for a solid-state system, and opened the way to a wealth of further theoretical studies. The basis functions adopted in the Yin-Cohen calculations were plane waves. All calculations were "static", the ions being fixed. Yin and Cohen considered 6 different crystal structures, and, for each of them, many different volumes (in Fig.12 the volume is measured in units of the experimental volume of bulk silicon). To obtain total energy and forces (forces can be obtained with the Hellmann-Feynman theorem: see Ihm et al. 1978, Nielsen and Martin 1984, Scheffler

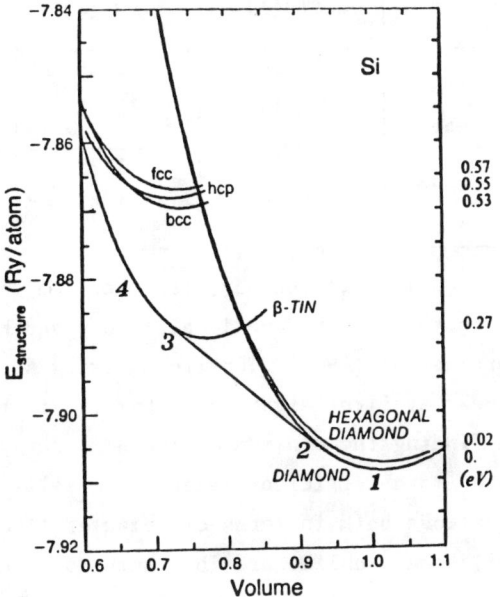

Fig.12 Total energy vs. volume for silicon in different crystal structures. (Yin and Cohen 1980, 1982)

et al. 1985) corresponding to each particular spatial configuration of ions, an independent self-consistent solution of the Kohn-Sham equations had to be sought from scratch. The results of Yin and Cohen for Si are reproduced in Fig. 12, where the calculated total energy is plotted against the cell volume for different crystal structures, and in Table I, which shows calculated versus experimental phonon data. The accuracy attained was impressive. The calculations obtained (i) the diamond structure as energetically most favorable (ii) the experimental equilibrium lattice constant, (iii) the experimental bulk modulus, (iv) some measured phonon energies, and finally even (v) the corresponding mode-Grüneisen parameters, which are related to third derivatives of the crystal energy with respect to atomic displacements. The only approximation here was LDA. All theoretical numbers were remarkably close to the experimental data.

Table I. Comparison with experiment of calculated frequencies ν (in THz) and mode-Grüneisen parameters γ for phonons at Γ and X in silicon.   (after Yin and Cohen 1982)

	LTO(Γ)	LOA(X)	TO(X)	TA(X)
ν_{theo}	15.16	12.16	13.48	4.45
ν_{expt}	15.53	12.32	13.90	4.49
γ_{theo}	0.9	1.3	0.9	-1.5
γ_{expt}	0.98	1.5	0.9	-1.4

Bulk silicon has also been the test case for the unified approach to Density Functional Theory and Molecular Dynamics of Car and Parrinello (1985). The two striking advantages of the method, which appear at first sight, are that (i) it represents an alternative way of finding the minimum of the HKS energy-density functional which is much more efficient than the self-consistent solution of the KS equations both in terms of computer time and of computer memory, and (ii) that, unlike previous methods which were intrinsically limited to static calculations (each new ionic configuration would require an independent self-consistent calculation), it allows the molecular-dynamics simulation: the motion of atoms in solids. In other words with the Car-Parrinello approach the electrons, after reaching their ground state (once and for all) for some initial ionic configuration, smoothly follow the ionic motion in configuration space along the Born-Oppenheimer surface. The latter ability, among other virtues, makes this method one of the most reliable tools to find, with the aid of annealing cycles, (i.e. by heating up and then slowly cooling down) equilibrium geometries of those complex systems (molecules or low-symmetry solid-state systems like defects or surfaces) where electron states cannot be neglected or lumped into some effective interatomic potential. Fig.13 shows one of the early results of this method: the real-time evolution of a small-amplitude distortion of the $\Gamma_{25'}$ symmetry in silicon (the $\Gamma_{25'}$ phonon).

Fig.13 Amplitude (top) and total electronic energy (bottom) vs. time for the $\Gamma_{25'}$ phonon in Si (Car and Parrinello 1985)

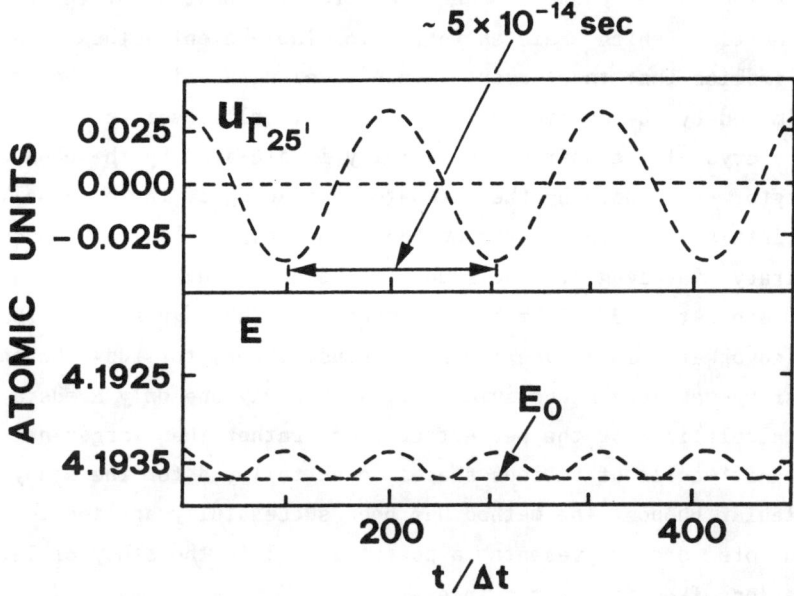

One will immediately notice the harmonic behavior of the crystal at low temperature (small displacement); the phonon frequency is read directly from the plot, and gives the same number which would be obtained from a static "frozen-phonon" calculation based on the same pseudopotential; the method proves completely successful.

The silicon test has opened the way to a variety of applications; presently finite-temperature properties of periodic systems with unit cells containing as many as 128 independent atoms can be studied from first principles, a task which ten years ago appeared to be beyond the most optimistic projections.

As far as phonons are concerned a recent development due to Baroni and coworkers (Baroni et. al 1987) should be mentioned. Traditionally, the way of studying phonons was either by calculating total energy and forces due to a particular phonon, which was obtained by slightly distorting the perfect crystal with the appropriate phonon wave, and then evaluating the frequency from the harmonic force-displacement (or energy-displacement) relation, or, equivalently, by choosing some small distortion of lower symmetry and then evaluating a few phonon frequencies from the portion of the dynamical matrix (which relates forces to displacements) thus obtained. This implies that the largest phonon wavelength which can be studied is limited by the maximum size of the unit cell, because in the distorted crystal the spatial periodicity is dictated by the phonon wavelength, and not by the interatomic spacing of the original perfect-lattice; in general the method is not applicable to phonons of arbitrary wavelength. The same limitation is encountered when phonons are studied via molecular dynamics. The suggestion of Baroni and coworkers is to use linear-response theory to study phonons of arbitrary wavelength and symmetry. In this way one only needs a single calculation for the perfect crystal, rather than larger and larger unit cells of various shapes, each tailored for the study of a particular phonon. The method has been successfully applied to semiconductors and represents a powerful tool in the study of lattice vibrations from first principles.

4.c. Transition metals

Norm-conserving pseudopotentials and the local-density approximation for exchange and correlation were also successfully used to investigate transition metals, a subject well outside the reach of previous, local pseudopotential theories. Pioneering results in this field were published by Ho, Harmon, Weber and Hamann (1982). In their work a mixed basis of gaussians and plane waves was adopted. Remarkable agreement with experiments was obtained for structural and vibrational properties of elemental Nb and Mo; moreover the calculations allowed a detailed analysis of the microscopic mechanisms causing phonon anomalies and soft-mode phase transitions in bcc Zr. Fig. 13 compares the calculated total energy per atom as a function of the displacement δ along the (1,1,1), i.e., the L phonon mode in

Nb, Mo, and bcc Zr; the potential energy curve corresponding to Zr, which had never been calculated previously along this important coordinate, nicely predicts the instability (known from experiments) of the bcc phase of Zr towards the formation of the omega phase, formed as two <111>

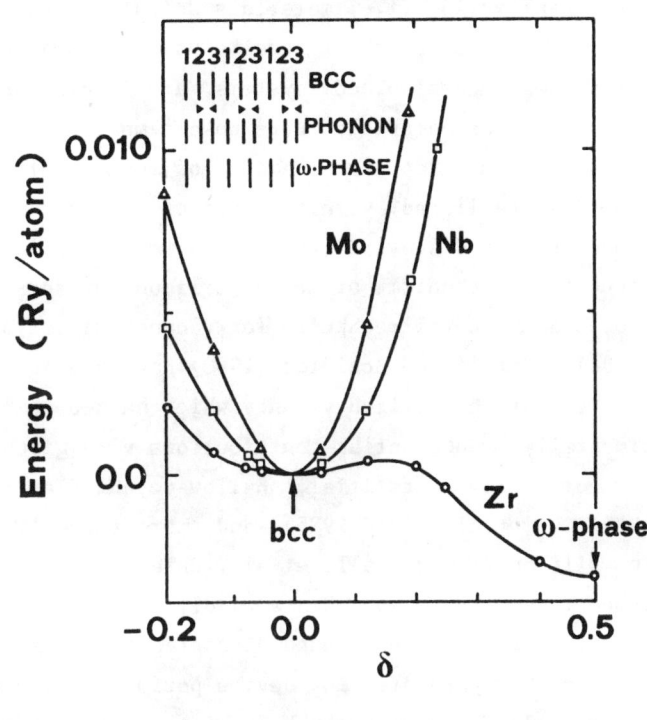

Fig.14 Total energy of Mo, Nb and bcc Zr vs. displacement δ (see inset) as obtained using the LDA and norm-conserving pseudopotentials (Ho et al.)

planes in the bcc unit cell collapse together (inset of Fig. 14).

4.d. A reliable microscope for low-symmetry systems ?

The fact that first-principles theories based on local-density functionals for exchange and correlation and norm-conserving pseudopotentials for the core-valence interaction reproduce to a high degree of accuracy experimentally known properties of crystals, shedding new light on their intepretation, has its own interest. However the main importance of such results is that they demonstrate the reliability and predictive power of the approach, and they encourage its use in the study of less known, low-symmetry solid-state systems, which are at the forefront of pure and applied research. The availability of a reliable microscopic theory has considerably increased the interaction of theoretical solid-state scientists with "the real world" (i.e. materials scientists and engineers); in just a few years a wealth of results was obtained on defects, surfaces, interfaces, chemisorbed systems, atomic clusters, and even liquids and amorphous materials. Here no attempt is made to give a complete account of this impressive amount of successful work, and only a few examples are listed. With the type of theory described in this paper Pandey (1982) discovered what is currently believed to be the reconstruction mechanism of clean silicon surfaces and, more recently, proposed a novel mechanism for vacancy migration in semiconductors (1986). Baraff and Schlüter (1980) have predicted "negative-U" properties of the silicon vacancy which have been later detected experimentally. Substantial contributions were given to the understanding of self-interstitial, shallow-dopant (Bar-Yam and Joannopoulos 1984, Car et al. 1984, 1985), and, more recently, hydrogen migration in silicon (Van de Walle et al., Chiarotti et al. 1988). Dielectric properties (Kunc and Resta 1983, Baroni and Resta 1986), and the physics of the valence-band discontinuity at semiconductor interfaces, a key parameter for device performance, were also microscopically understood with the help of this kind of theory (Van de Walle and Martin 1987, Baldereschi et al. 1988, Christensen 1988). Finally, the unified approach to Molecular Dynamics and DFT has opened the way to the study of complex systems like amorphous silicon and carbon (Car and Parrinello 1988, Galli et al. 1988) and small atomic clusters (Hohl et al. 1987, Ballone et al. 1988, Andreoni 1988).

After this (incomplete) list of glories, it may be fair to list a number of limitations of local-density functionals. (i) "overbinding": absolute binding energies are overestimated, although relative energies are very accurate (Jones 1984) (ii) "self-interaction": electron affinities are out of reach (Perdew and Zunger 1981) (iii) "gaps": excited states in insulating solids are not trivially related to ground-state properties, their description requiring much additional work (Hybertsen and Louie 1985, Godby et al. 1986) (iv) "3d transition metals": s-d transfer energies of are not well described (Gunnarsson and Jones 1985) (v) highly correlated electrons, which seem to be at the heart of the high-T_c superconductors, are also out of reach. Presently it seems that any extension which attempts to cure one of these problems either spoils something else, or becomes so involved as to loose much of the appeal of local density-functional approaches, which rests in their remarkable simplicity. Fortunately, a lot of physics and chemistry of materials can be safely studied within these known limits. But the new interest of solid-state theorists towards Quantum Monte Carlo methods suggests that some open problems may soon be attacked from an entirely different side.

5. Conclusions

Theoretical solid-state physics and theoretical chemistry are getting closer, as far as first-principles approaches are concerned; occasions of scientific exchange are highly valuable. The intention of this overview was to give some idea of the recent developments of the theory of electronic states in the physics community, and to shed some light on the future of the field. The discovery and successful application of norm-conserving pseudopotentials has represented a major step forward in the first half of the eighties. The unification of Molecular Dynamics and Density Functional Theory, on one hand, and the direct simulation of quantum many-body systems (with appropriate extensions of pseudopotential theory), on the other, appear as promising directions for this second half of the eighties. Artificial intelligence and symbolic manipulation, not discussed here, may also play some role. In general it seems that the venerable, old branch of electronic structure, far from shrinking into a brute-force number-crunching arena of declining scientific interest, represents even today a challenging and lively meeting place for many-body physicists, chemists, and materials scientists.

References

ALMBLADH   C.-O. and von Barth U., Phys. Rev. B $\underline{31}$,3231 (1985)
ALDER   B.J. and Wainwright T.E., J. Chem. Phys. $\underline{33}$, 1439 (1960)
ANDERSEN   O.K., Phys. Rev. B $\underline{12}$, 3060 (1975)
ANDERSON   J.B., J. Chem. Phys. $\underline{63}$, 1499 (1975)
ANDREONI   W., Proceedings of the NATO/ASI Workshop on "Semiconductor Microstructures and Band-Structure Engineering" (Lucca 1988, R.A. Abram, Ed.), in print
ANTONCIK   E., J. Phys. Chem Solids $\underline{10}$, 314 (1959)
APPELBAUM   J.A. and Hamann D.R., Rev. Mod. Phys. $\underline{48}$, 3 (1976)
ASHCROFT   N.W. and Mermin N.D., "Solid State Physics". Holt-Saunders International, New York 1976
BACHELET   G.B. and Schlüter M., Phys. Rev. B $\underline{25}$, 2103 (1982)
BACHELET   G.B., Hamann D.R. and Schlüter M., Phys. Rev. B $\underline{26}$, 4199 (1982)
BACHELET   G.B., and Christensen N.E., Phys. Rev. B $\underline{31}$, 879 (1985)
BACHELET   G.B., in "Crystalline Semiconducting Materials and Devices" (P.N. Butcher, N.H. March and M.P. Tosi, Eds.), Plenum 1986
BACHELET   G.B., Ceperley D.M. and Chiocchetti M.G.B., preprint (1988)
BALDERESCHI   A., Phys. Rev. B $\underline{7}$, 5212 (1973)
BALDERESCHI   A., Resta R. and Baroni S., preprint (1988)
BALLONE   P., Andreoni W., Car R. and Parrinello M., Phys. Rev. Lett. $\underline{60}$, 271 (1988)
BARAFF   G.A., Kane E.O. and Schlüter M., Phys. Rev. B $\underline{21}$, 3563 and 5662 (1980)
BARONI   S. and Resta R., Phys. Rev. B $\underline{33}$, 7017 (1986)
BARONI   S., Giannozzi P. and Testa A., Phys. Rev. Lett. $\underline{58}$, 1861 (1987); $\underline{59}$, 1662 (1987)
BAR-YAM   Y. and Joannopoulos J.D., Phys. Rev. Lett. $\underline{52}$, 1129 (1984)
BASSANI   F. and Pastori Parravicini G., "Electronic States and Optical Transitions in Solids", Pergamon 1974

BERNARDES   N., Phys. Rev. 112, 1534 (1958)
BINDER   K. (ed.): "Monte Carlo methods in statistical physics", Springer 1979
BINDER   K. (ed.): "Monte Carlo methods in statistical physics II", Springer 1984
CALLAWAY   J., J. Math. Phys. 5, 783 (1964)
CAR   R., Kelly P.J., Oshiyama A. and Pantelides S.T., Phys. Rev. Lett. 52, 1814 (1984); ibid. 54, 360 (1985)
CAR   R. and Parrinello M., Phys. Rev. Lett. 55, 2471 (1985)
CAR   R. and Parrinello M., Phys. Rev. Lett. 60, 204 (1988)
CAR   R. and Parrinello M., preprint (1988)
CEPERLEY   D.M. and Kalos M.H., in: Kalos 1979
CEPERLEY   D.M. and Alder B.J., Phys. Rev. Lett. 45, 566 (1980)
CEPERLEY   D.M. and Alder B.J., J. Chem. Phys. 81, 5833 (1984)
CEPERLEY   D.M. and Alder B.J., Phys. Rev. B 36, 2092 (1987)
CERNY   V., preprint, Comenius University, Bratislava 1982
CHELIKOWSKY J.R. and Cohen M.L., Phys. Rev. B 14, 556 (1976)
CHIAROTTI   G., Buda F., Car R. and Parrinello M., to be published
CHRISTENSEN N.E., preprint (1988)
CHRISTIANSEN P.A., Lee Y.S. and Pitzer K.S., J. Chem. Phys. 71, 4445 (1979)
COHEN   M.L., Physics Today, p. 40 (July 1979)
GALLI   G., Car R. and Parrinello M., to be published
GYGI   F. and Baldereschi A., Phys. Rev. B 34, 4405 (1986)
GODBY   R.W., Schlüter M. and Sham L.J., Phys. Rev. Lett. 56, 2415 (1986)
GUNNARSSON   O. and Lundqvist B.I., Phys. Rev. B 13, 4274 (1976)
GUNNARSSON   O., Jonson M. and Lundqvist B.I., Phys. Rev. B 20, 3136 (1979)
GUNNARSSON   O. and Jones R.O., Physica Scripta 21, 394 (1980)
GUNNARSSON   O. and Jones R.O. in "Local Density Approximations in Quantum Chemistry and Solid State Physics" (J.P. Dahl, J. Avery, Eds.), Plenum 1984

GUNNARSSON O. and Jones R.O., Phys. Rev. B 31, 7588 (1985)
HAMANN D.R., Phys. Rev. Lett. 42, 662 (1979) and unpublished
HAMANN D.R., Schlüter M. and Chiang C., Phys. Rev. Lett. 43, 1494 (1979)
HARMON B.N., Weber W. and Hamann D.R., Phys. Rev. B 25, 1109 (1982)
HARRIS J. and Jones R.O., Phys. Rev. Lett. 41, 191 (1978)
HEDIN L. and Lundqvist B.I., J.Phys.C 4, 2064 (1971)
HEINE V., Cohen M.L. and Weaire D., "Solid State Physics", vol.24 (H.E. Ehrenreich, F. Seitz and D. Turnbull, Eds.), Academic 1970
HO K.-M., Fu C.-L., Harmon B.N., Weber W., and Hamann D.R., Phys. Rev. Lett. 49, 673 (1982); Fu C.-L. and Ho K.-M., Phys. Rev. B 28, 5480 (1983); Ho K.-M., Fu C.-L., Harmon B.N., ibid. 6687
HOHENBERG P.C. and Kohn W., Phys. Rev. 136, B864 (1964)
HOHL D., Jones R.O., Car R. and Parrinello M., Chem. Phys. Lett. 139, 540 (1987)
HYBERTSEN M.S. and Louie S.G., Phys. Rev. Lett. 55, 1418 (1985)
IHM J., Zunger A. and and Cohen M.L., J. Phys. C 12, 4409 (1979)
IHM J. and Cohen M.L., Sol. State Commun. 29, 711 (1979)
JACUCCI G., in "Diffusion in Crystalline Solids" (G.E. Murch and A.S. Norwick, Eds.), Academic 1985
JOANNOPOULOS J.D., Starkloff Th. and Kastner M., Phys. Rev. Lett. 38, 660 (1977)
JONES R.O., Phys. Rev. Lett. 52, 2002 (1984)
JONES R.O., Phys. Rev. A 32, 2589 (1985); J. Chem. Phys. 82, 5078 (1985); ibid. 84, 318 (1985)
JONES R.O., Chem. Phys. Lett. 125, 221 (1986)
KAHN L.R. Goddard W.A., J. Chem. Phys. 56, 2685 (1972)
KALOS M.H., Levesque D. and Verlet L., Phys. Rev. A 9, 2178 (1974)
KALOS M.H. (ed.), "Monte Carlo Methods in Quantum Problems" NATO/ASI Series C, Reidel 1984
KIRKPATRICK S., Gelatt G.D.,jr., and Vecchi, M.P., Science 220, 671

	(1983)
KLEINMAN	L., Phys. Rev. B $\underline{21}$, 2630 (1980)
KOHN	W. and Sham L.J., Phys. Rev. $\underline{140}$, A1133 (1965)
KOHN	W., Phys. Rev. Lett. $\underline{51}$, 1596 (1983)
KUNC	K. and Resta R., Phys. Rev. Lett. $\underline{51}$, 686 (1983)
LANDAU	L. and Lifschitz E., "Mecanique Quantique", Mir, Moscow 1966
LOUIE	S.G., preprint (1988)
MARADUDIN	A.A.,"Solid State Physics", vols. 18,19 (F. Seitz and D. Turnbull, Eds.), Academic 1966
MELIUS	C.F. and Goddard W.A., Phys. Rev. A $\underline{10}$, 1528 (1974)
MERMIN	N.D., Phys. Rev. $\underline{137A}$, 1441 (1965)
METROPOLIS	N., Rosenbluth A.W., Rosenbluth M.N., Teller A.M., and Teller E., J. Chem. Phys. $\underline{21}$, 1087 (1953)
MITAS	L., Bachelet G.B. and Ceperley D.M., in print (1988)
MOSKOWITZ	J.W., Schmidt K.E., Lee M.A., and Kalos M.H., J. Chem. Phys. $\underline{77}$, 349 (1982)
NIELSEN	O.H. and Martin R.M., Phys. Rev. Lett. $\underline{50}$, 697 (1983)
OLIVEIRA	L.N., Gross E.K.U. and Kohn W., Phys. Rev. Lett. $\underline{60}$, 2430 (1988)
PANDEY	K.C., Phys. Rev. Lett. $\underline{49}$, 223 (1982)
PANDEY	K.C., Phys. Rev. Lett. $\underline{57}$, 2287 (1986)
PERDEW	J.P. and Zunger A., Phys. Rev. B $\underline{23}$, 5048 (1981)
PHILLIPS	J.C. and Kleinman L., Phys. Rev. $\underline{116}$, 287 (1959)
RAHMAN	A., Phys. Rev. $\underline{136}$, A405 (1964)
REDONDO	A., Goddard W.A. and McGill T.C., Phys. Rev. B $\underline{15}$, 5038 (1977)
REYNOLDS	P.J., Ceperley D.M., Alder B.J., and Lester W.A., J. Chem. Phys. $\underline{77}$, 5593 (1982)
SCHEFFLER	M., Vigneron J.P. and Bachelet G.B., Phys. Rev. B $\underline{31}$, 6541 (1985)
SCHMIDT	K.E. and Kalos M.H., in: Binder 1984
SHAW	R.W. and Harrison W.A., Phys. Rev. $\underline{163}$, 604 (1967)
SINGWI	K.S., Tosi M.P., Land R.H., and Sjölander A., Phys. Rev. B

	$\underline{1}$, 1044 (1970)
STARKLOFF	Th. and Joannopoulos J.D., Phys. Rev. B $\underline{16}$, 5212 (1977)
STICH	I., Car R., Parrinello M. and Baroni S., preprint (1988)
TOPIOL	S., Zunger A. and Ratner M.A., Chem. Phys. Lett. $\underline{49}$, 367 (1977)
TOPP	W.C. and Hopfield J.J., Phys. Rev. B $\underline{7}$, 1295 (1974)
TOSI	M.P., "Solid State Physics", vol.16 (F. Seitz and D. Turnbull, Eds.), Academic 1964
VAN DE WALLE	C., Bar-Yam Y. and Pantelides S., preprint (1988)
VAN DE WALLE	C. and Martin R.M., Phys. Rev. B $\underline{35}$, 8154 (1987)
VERLET	L., Phys. Rev. $\underline{159}$, 98 (1967)
WENDEL	H. and Martin R.M., Phys. Rev. B $\underline{19}$, 5251 (1979)
YIN	M.T., in "Proceedings of the 17th ICPS" (D.J. Chadi and W.A. Harrison, Eds.), Springer 1985
YIN	M.T. and Cohen M.L., Phys. Rev. Lett. $\underline{45}$, 1004 (1980)
YIN	M.T. and Cohen M.L., Phys. Rev. B $\underline{26}$, 5668 and 3259 (1982)
ZIMAN	J.M., "Principles of the Theory of Solids", Cambridge University Press 1969
ZUNGER	A. and Cohen M.L., Phys. Rev. B $\underline{18}$, 5449 (1978)

Autodeductive Modeling and Optimization in Chemometrics

Mario Marsili

Computer Chemistry Laboratory
Chemistry Department
Universita' de L'Aquila
Via Assergi 4
67100 L'Aquila, ITALY

Introduction

The average chemist in academic and especially in industrial environment still considers computers and software as tools reserved only for selected specialists. Our experience in interacting with chemical, pharmaceutical and car industry within a global research project in Computer Chemistry [1,2] allows to understand the essential demands regarding experimentation, modeling and optimization. In general, highly redundant measurements are carried out; in most cases there is a lack of deterministic equations describing the physics of the experiment, and data appear to contain much noise. However, it seems sometimes difficult to convince people to involve computer support due to a certain difficulty in interpreting results of programs showing a highly mathematical layout. The average chemist prefers undoubtedly a more semantic, colloquial access to a computer session . This is reflected in the actual trend of creating *thinking* programs, like the *expert systems*, which emulate human logic and are used mostly for *problem diagnostics*. A central point in industrial experimentation is *optimization*. Optimizing an experiment means reducing costs and times; further it leads to a better planning for the future . Optimization is linked to a specific *best* operational strategy, which has to be found.

To optimize an experiment one has to select particular values for only those independent variables which really model the response(s) of the experiment under study, and reject all other non significant variables . This is equivalent to separating information from noise. Elimination of non-relevant variables, as e.g. attempted by the classical *stepwise multilinear regression analysis* , leads already to an optimization: the experimenter ends up with the minimal number of parameters sufficient to model the maximum information within his observed system. Each eliminated independent variable means one less operation in real laboratory work .

In elemental geological analysis, for example, reducing the number of ions to be determined in order to classify unknown samples directly optimizes the overall analytical performance .

In other cases a model, a certain equation describing the data, is desirable in order to find the set of parameter values that correspond to a maximum in the experimental response. If deterministic equations are available then standard methods can be applied [3] to find an optimum. We deal here with situations where such a deterministic function is unavailable, a common situation in many industrial processes : some statistical model must be derived instead.

We pursued the double goal of creating a software package, SPECTRE-M, that would offer a new, powerful method in computer-assisted modeling and optimization of multivariate experiments and allow an easy human-computer interaction even to non-professional chemometricians. The SPECTRE-M system contains a self-deciding part, based on empirical and statistical rules, that provides for the *autodeductive* features of the system [4].

Current Methods: Limits and Benefits

It is often useful to establish some statistical model relating independent variables x (experimental parameters) to dependent variables y (measured responses). A good model (which can be obtained only from a good experimental planning) is the necessary key to any optimization. However not every statistical modeling algorithm is equally suitable for modeling and prediction. For example, multilinear regression analysis, MLR, although providing the best method for *fitting* data, may not be useful for *predicting* new data if one or more of the following cases holds:
1) the dependent variables contain much noise

2) the independent variables are colinear
3) there are more *x* variables than data vectors
4) the data are clustered
5) simultaneous prediction of more than one related responses.

The advantage of MLR is clearly its directness of interpretation. The independent variables appear in the final regression equation (eq.1) along with their

$$y = \Sigma c_i x_i \qquad \text{eq.1}$$

importance, i.e. the regression coefficients c_i. In reality almost every response is a combination of information and noise. To separate them, a certain response matrix Y (a vector y for single-response experiments) can be represented by a *principal component decomposition* according to the PLS algorithm [5][6]. PLS seems at the moment the best method to model a block of response data from a block of independent variables as it eliminates all the shortcomings of MLR listed above. The noise, which is included in the MLR model, becomes separated in a PLS formalism and confined on a *residuals* matrix R_Y. The information is contained in *rotated principal components* of the X-block and of the Y-block, the *latent variables*, computed such as to maximize the variance explained in the response matrix Y.

Therefore an *inner relation* links both PC-spaces by eq.2

eq.2 $$\mathbf{u}_j = b_j \mathbf{t}_j$$

where u_j are the the scores on the j-th latent variable in the Y space, and t_j are those of the j-th latent variable in the X space. The latent variables themselves are given by following equations

eqs. 3a-b
$$X = \sum t_j p_j + R_X \quad ; \quad Y = \sum u_j q_j + R_Y$$

The above equations, called *outer relations*, say that the two original matrices can be thought as of consisting of a sum of j rank-one matrices.

The matrix elements are the results from the outer products of scores (t and u) and loadings (p and q), plus full-rank matrices R containing the respective residuals (the error matrices). The goal is to have the residuals matrix *as small as possible*. This means philosophically that a *best predictive model*, and not a best *fitting* model (like in MLR) must be found. To understand the difference between a classical Principal Components Regression and the latent variables approach we first discuss the formalism of the first method in a NIPALS algorithm (Non Iterative Partial Least Squares). NIPALS is sustantially a method for eigenvalue extraction.

The method starts taking a vector x_i out of X as temptative first principal component.

eq. 4
$$t_{j(start)} = x_i$$

$$p_j^t = t_j^t X (t_j^t t_j)^{-1}$$

and after normalization of the loadings

eq. 5
$$t_j = Xp(p_j^t p_j)^{-1}$$

the scores can be computed. This procedure proceeds till convergence. With he same equations other principal components are calculated for the Y block: in PCR the various PC(y) and PC(x) are regressed in the sense of eq. 2 to find the best correlation, as shown in eq. 6

eq. 6
$$b_j = u_j^t t_j (t_j^t t_j)^{-1}$$

So far, the Y block and the X block have been processed separately.

In the PLS formalism, however, the above equations are changed as follows

$$u_j = \text{any } y_i \text{ in original matrix}$$

eq. 7
$$p_j^t = u_j^t X (u_j^t u_j)^{-1}$$

normalize p and then

eq. 8
$$t_j = X p_j^t (p_j^t p_j)^{-1}$$

and in analogy

eq. 9
$$q_j^t = t_j^t Y (t_j^t t_j)^{-1}$$

and after normalization

eq. 10
$$u_j = Y q_j (q_j^t q_j)^{-1}$$

The equations 7-10 have all least-squares structure like

eq. 4 *but* **t** *and* **u** *have exchanged place.* Thus, it is empirically achieved that the **X** and **Y** worlds know about each other, leading to the PLS typical rotated eigenvectors.

Now we can finally write eq. 11 *(for each component)*

eq. 11 $$\mathbf{R}_y = \mathbf{Y} - \mathbf{u}q$$

and using the inner relation

eq. 12 $$\mathbf{R}_y = \mathbf{Y} - b\mathbf{t}q$$

The regression coefficients **b** are again calculated by eq. 6.

The maximum number of latent variables in a PLS model is obviously equal to the number of independent variables in **X**. However, only the a number of latent variables best apt to predict **Y** with the lowest error, i.e. with the minimal Predicted Residual Sum of Squares (PRESS) is significant.

The definition of this minimal number is obtained by the *cross-validation* method, which is of central importance in the SPECTRE-M autodeductive evaluation of the importance of *independent variables.* The cross-validation technique works on a reduced set of objects in **X**. For example, every third data vector is left out of **X** and the smaller matrix **X'** is processed by the PLS algorithm. The remnant third of data vectors, the *test set*, are *predicted* by a PLS model augmented each time by one dimension with respect to the preceding model. The dimensionality of the PLS model constructed on **X'** which leads to the smallest error matrix **R** is considered as the best predictive, lowest-rank model.

The method can efficiently separate intrinsic noise from

information, but we still face the strategic problem of the variables' importance. The PLS model generated with the discussed algorithm still contains all original independent variables.

A straightforward interpretation of the *global importance*, i.e. of the strategic role of each x variable is more difficult to do. If on the one hand PLS is capable of revealing what is noise and what is information, thus giving better predictive models, on the other hand it cannot eliminate another important but different kind of noise: the presence of variables in the model that do not contribute to explain Y, the *non-significant* variables. Elimination of such variables is a strategic action as it results immediately in an *operational optimization of the experimental set-up*. In other words, if a certain variable, say the temperature, is found not to be necessary for describing some measured response(s), one can exclude *temperature* from the experiment, avoiding costs coming from heating equipment, control devices and so on. In addition a refined model is obtained which could be closer to the *physics* of the experiment studied. A response optimization is also a direct consequence of generating a model largely filtered from these two sources of noise. Here we refer to the classic semantic of *optimization:* to obtain a maximum response. Only through a refined, noise-free mathematical model one can predict the maximum of the related response hypersurface.

The SPECTRE Evaluation and Sensitivity Method

Within SPECTRE a possibility of operational and of classical optimization is offered by its autodeductive evaluation system . It is a unique powerful synthesis of the MLR and PLS philosophies. This system uses:
1- a PRESS-driven PLS approach to eliminate intrinsic noise in the responses, leading to a model of lower dimensionality by deletion of latent variables covering noise
2- Classification of original independent variables into *significant or non-significant* , by a combination of an iterative *leave-one-out* mechanism supported by extended cross-validation techniques. Variables that worsen the prediction quality are considered as *non-significant* to the model. The cross-validated model with a combination of independent variables yielding the *lowest* PRESS, i.e. maximizing the *variance explained in* Y is considered as the best refined predictive model.
3- Ranking of the non-significant variables x_{non} by a *sensitivity analysis* of the current PLS model sensitivities $s(X_k)$ to a given standard *perturbation* p_k introduced on each variable X_k in turn.
$$s(X_k) = [F(x_1, x_2 \ldots x_k \ldots x_n) - F(x_1, x_2, \ldots, x_k + p_k, \ldots x_n)]/ \sigma(X_k)$$
4- Elimination of non-significant variables beginning with

the one with the lowest sensitivity value.

5- Transformation of sensitivities of the survived variables of a *PLS model* into *MLR*-like *coefficients* by multiplication of each $s(x_k)$ by the variance $\sigma(x_k)$.

The great advantage of SPECTRE-M is to deliver a refined model in a single equation. Thus, the role of the survived variables becomes immediately understandable, in contrast to the loadings representation of the importance of a variable in a traditional PLS approach. Nevertheless the SPECTRE-M method is reliable for predicting test sets, as it does not suffer from the MLR (or stepwise MLR) deficiencies, because it relies on a sound cross-validated (i.e. trained on prediction) functional framework. The final model can now be used for detection of a maximum of the response(s). Two methods are currently available: a grid search algorithm and a *simplex* search method. The user, once that the best predictive function has been obtained, has to define interactively the ranges of validity of each survived independent variable to provide for the boundaries of the search area. In the grid search method a net of points is computed at given established intervals on the response hypersurface. Around the points showing highest values the grid is tightened and the intervals become smaller. The procedure ends when all relative maxima and the absolute maximum have been localized.

The *simplex* method can be applied anytime during the

search session when the number of relative maxima is small. The starting points for the *simplex* are determined by the edges of the hyperpolygon on the grid that shows the highest values.

We shall illustrate the performance of SPECTRE-M on a simple synthetic example of reactor optimization.

Example of SPECTRE-M Optimization

The performance of a chemical reactor should be maximized. The independent variables describing the operational mode of the system are : temperature (°C), pressure (bar), and stirrer speed (rpm). The dependent variables, the responses, are : yield (gr) and cost ($) for each experiment.

The data set is shown on Table 1.

Temperature	Pressure	Stirrer	Yield	Cost
25.13	3.83	0	2503.3	35.9
25.68	7.16	15	2523.6	67.8
78.44	8.82	320	1769.2	248.2
11.06	1.00	120	1729.3	65.6
55.34	8.03	320	2785.5	224.2
26.72	0.80	120	2544.8	78.0
4.79	1.00	100	1264.5	52.5
69.94	5.74	360	2277.6	225.8
49.71	4.73	320	2877.4	190.3
33.00	0.50	0	2732.6	34.4
92.95	7.48	60	595.8	138.0
14.76	4.42	60	1984.9	56.5

Table 1 Reactor example data set

Two questions are interesting : first, which is the best predictive model relating the independent to the dependent block of variables? And second, what are the experimental conditions that guarantee maximum yield at *lowest* costs? This last question requires a double optimization and can be reformulated as follows: what is the maximum obtainable overall gain for the reactor experiment? We define therefore the variable 'gain' by eq.13

eq.13   gain = yield*(value of 1 gr of product) - cost

The original variables x were first expanded to include non-linear contributions (quadratic and interaction terms). Thus a matrix of nine independent variables was obtained ($T, T^2, p, p^2, s, s^2, Tp, Ts, ps$).

REFINED MODEL COEFFICIENTS :

INTERCEPTS =   846.5039   8.3145

X	1 (temp	)	FOR Y1: 0.8999E+02 FOR Y2: 0.8290E+00
X	2 (pressure	)	FOR Y1: 0.1086E+02 FOR Y2: -.8830E+00
X	3 (stirrer	)	FOR Y1: -.6104E-03 FOR Y2: 0.4039E+00
X	4 (temp ** 2	)	FOR Y1: -.9999E+00 FOR Y2: -.1058E-02
X	5 (pressure ** 2	)	FOR Y1: -.1018E+01 FOR Y2: 0.7765E+00

TABLE 2. THE FINAL MODEL EQUATIONS

After elimination of some non-significant variables the model equations of Table 2 were obtained for yield ($y1$) and cost ($y2$).

The functions for $y1$ and $y2$ are merged to give an expression according to eq.13, which can now be globally optimized.

$$F(gain) = yield*0.031 - cost$$

being 0.031$ assumed as the raw price of 1 gr of product. Table 3 shows the results listing the first five best combinations of values for T, p, and s. At the top appears the setting giving the maximum performance (highest yield at lowest price).

As interesting comparison the same data set has been processed with a stepwise MLR algorithm available in the chemometrical software package ARTHUR [7]. A different model function was computed for the yield (see Table 4)

It appears that out of the $n!/m!(n-m)!$ possible ways of selecting m variables one combination was found that provides for a good data fit. However, as shown in the column refering to the prediction errors, the noise present in the raw data is carried along in the fitting mechanism of the SMLR procedure. This causes the predictions to be less accurate than in the SPECTE-M approach.

NR.	GAIN	YIELD	COST	Temp	Press	Stirrer
1	50.27	2781.10	35.94	35.00	1.00	0.00
2	50.21	2656.13	32.13	30.00	1.00	0.00
3	50.16	2782.82	36.10	35.00	1.20	0.00
4	50.09	2657.86	32.30	30.00	1.20	0.00
5	49.99	2784.47	36.33	35.00	1.40	0.00

TABLE 3. THE FIVE BEST SETTINGS FOR THE VARIABLES GOVERNING THE REACTORS PERFORMANCE

Conclusions

SPECTRE-M provides a useful help to the experimentalist facing problems of optimization (strategical and tactical). The straightforward interpretability of the mathematical models allows immediate perception of the importance of the various features governing an experiment.

SPECTRE-M is a software system written in FTN 77 and supported by a number of graphic output routines, that help in the visual comprehension of the results.

Currently it is being integrated into the more general SPECTRE project, aiming at creating a chemometrical expert system of wide utility to the industrial chemist [8].

PREDICTIONS:

Temp	Press	Stirrer	Y pred.(1)	Y pred.(2)	Y true	Err.(1)	Err.(2)
25.0000	1.0000	60.0000	2481.1	2483.9	2480.4	+ 0.7	+ 3.5
40.0000	2.0000	60.0000	2863.8	2861.0	2862.9	+ 0.9	− 1.9
60.0000	1.0000	120.0000	2655.8	2657.3	2654.7	+ 1.1	+ 2.6
10.0000	10.0000	0.0000	1653.2	1680.8	1652.9	+ 0.3	+27.9
80.0000	10.0000	120.0000	1652.7	1677.3	1651.5	+ 0.6	+25.8
35.0000	5.0000	0.0000	2800.1	2793.4	2799.1	+ 1.0	− 5.7
70.0000	2.0000	180.0000	2263.5	2258.9	2262.3	+ 1.2	− 3.4
40.0000	5.0000	60.0000	2875.0	2868.3	2874.0	+ 1.0	− 5.7

(1) Data predicted by SPECTRE-M
(2) Data predicted by STEP from "ARTHUR"

$Y = 90.00\ T + 2.44\ P - 1.00\ T^{**}2 + 856.1$ (function obtained by STEP)

TABLE 4

References

[1] M. Marsili, H. Saller, E. Marengo and M. Salomone
     Chimica Oggi, 12 (1985) p.57

[2] M. Marsili, E. Marengo, M. Salomone, C. d. Buono,
     F. Cammarata, G. Scavia and L. Caglioti
     Chimica e Industria, 5 (1987)

[3] see for example
     a) J. Kuester and J. H. Mize, in 'Optimization Techniques
     with FORTRAN' , McGraw-Hill, 1973
     b) D. L. Massart, A. Dijkstra and L. Kaufman, in
     'Evaluation and Optimization of Laboratory Methods and
     Analytical Procedures' , Elsevier, Amsterdam, 1984

[4] M. Marsili, E. Marengo and H. Saller, Anal. Chim. Acta,
     in press

[5] S. Wold, A. Ruhe, H. Wold and W. J. Dunn III, SIAM
     J. Stat. Comp., 5 , 1985, p.735

[6] B. Kowalski Ed., Chemometrics: Mathematics and
     Statistics in Chemistry Reidel Publ., Dordrecht
     1984, pp.17-95

[7] B. Kowalski   Ed., Chemometrics: Theory and
     Application, ACS Symp. Ser. 52 , Washington 1977

[8] M. Marsili
     Chimica e Industria, 4, 1988

Statistical Distribution of Molecular Conformations and Its Application in QSAR Research.

Raimondo Scordamaglia, Luisa Barino

Istituto G. Donegani,
Via G. Fauser 4, I-28100 Novara

Abstract

The CSD (Conformation Statistical Distribution) method is presented. It carries out a very fast scanning of the whole conformational hypersurface of a molecule having n internal rotational degrees of freedom, through the calculation of the nonbonded intramolecular energy for each point of a n-dimensional homogeneous grid in the surface. The method allows then to calculate both the probabilities (according to Boltzmann's statistics) with which the molecule assumes its proper energetically possible conformations and the statistical weights of the various conformational minima in which the molecule can be found. An averaged picture of the conformational possibilities of a bioactive molecule while approaching the receptor site can thus be given.
Some examples show the use of such a weighted mean in QSAR research aimed at the determination of molecular descriptors (conformational freedom, molecular shape, total dipole moment and other steric and electronic features) related to activities in series of bioactive compounds.

Introduction.

When studying the correlation between biological activity and molecular structure in series of chemical compounds, or when performing the conformational analysis of a single molecule to point out its steric features, it is important to determine not only the molecular minimum energy conformation but also the other energetically possible conformations and the statistical weights of the various conformational minima in which the molecule can be found. It is truly improbable that the molecule acts on the receptor in its minimum energy conformation and not in one of higher energy or rather in an excited state.

Thus, determination of the most probable conformational possibilities of a molecule while approaching the receptor has been our goal in structure-activity relationship studies. From the steric and electric properties of most active compounds in their most probable conformations, one can then try to extract those molecular features common to the best compounds and expressible as quantitative parameters (descriptors) or practical rules. This set of descriptors can correlate with activity values and can be calculated (or verified) for a new compound to estimate its properties. They might thus lead to the definition of a new compound, as maximum goal: more frequently they suggest, in a series of compounds, the substituents able to increase activity thus performing what is known as lead optimization.

Method.

The CSD method (Conformation Statistical Distribution) carries out a very fast scanning of the whole conformational hypersurface of a molecule having n internal rotational degrees of freedom, through the calculation of the non-bonded intramolecular energy for each point of a n-dimensional homogeneous grid in the surface. This method is aimed at the selection of the most probable conformation and of the statistical probabilities of all the other energetically allowed but less probable conformations. It gives a weighted average of the conformational possibilities of a molecule while approaching the receptor.

The goal of calculating all allowed conformations and partitioning them into well defined property clusters has been reached through original and compact algorithms, which led to three computer programs consequent one to the other, each of them aimed at one step of procedure as follows:

ERHYCA | (Entire Rotational Hypersurface Conformation Analysis) calculates geometry and non-bonded energy (through a Lennard-Jones 6-12 function plus the electrostatic term) of conformations generated by rotating the molecular portions around free internal rotation bonds with chosen-by-user step and under a high chosen energy threshold. Initial molecular geometry may be given as orthogonal coordinates of atoms or may be generated through the molecular structure generating program STERIMOL, which also provides topological matrix. [1] [2]
Absolute minimum under the above approximations can be calculated.

BOLSTAT

(BOLtzmann STATistics) calculates the statistical distribution function of conformation on discrete energy levels, i.e. the conformational partition function:

$$Z = \sum_i M_i \cdot e^{-\frac{1}{RT} \cdot E_i}$$

- E_i = ENERGY OF LEVEL i (Kcal/mole)
- n_i = N° CONFORMATIONS ON LEVEL i
- T = TEMPERATURE (°K)
- R = BOLTZMANN'S FACTOR (1.98×10^{-3} Kcal/mole*°K)

and allows the choice of the energy value under which to scan the rotational hypersurface from ERHYCA. Usually in calculations we take, at each T, the energy level having a probability of 0.1% as the max energy level (this probability value proved to be sufficient to allow the subsequent detection of all relative minima in the calculated hypersurface).

(CLUster DIStribution) defines in the hypersurface ensembles of conformations (clusters) separated by rotational energy barriers higher than the energy threshold and corresponding to all relative conformational minima under the approximations of the method. Shape and population of clusters changes with temperature and cluster pathways are defined.
Along these pathways the following quantities are calculated:

CLUDIS

- GEOMETRICAL LIMITS OF CLUSTERS as angular variation ranges of rotatable bonds

- PARTITION FUNCTION INSIDE CLUSTERS:

$$Z_j = \sum_i n_{ji} \cdot e^{-\frac{1}{RT} \cdot E_i}$$

$\begin{cases} j = \text{CLUSTER INDEX} \\ n_{ji} = \text{N° OF CONFORMATIONS ON LEVEL i IN CLUSTER j} \end{cases}$

- STATISTICAL WEIGHT OF CLUSTERS IN THE HYPERSURFACE:

$$P_j = Z_j / Z$$

- ENTROPY (GIBBS):

$$S_j = -R \sum_i p_{ji} \cdot \ln p_{ji}$$

$\begin{cases} \text{where } P_{ji} = \left(n_{ji} \cdot e^{-\frac{1}{RT} \cdot E_i} \right) / Z \\ \text{IS THE PROBABILITY OF LEVEL i IN CLUSTER j} \end{cases}$

Examples.

As an example of application, we report the results obtained in the search for common features explaining activity for a certain number of N-phenylamides used as systemic fungicides, particularly in the control of species Peronosporales. The full results have been presented elsewhere [3]. Studied compounds are shown in Fig. 1.

From the steric characteristics of most active compounds in their most probable conformations we have found that the two central carbonyl groups (one linked to nitrogen and the other contained in substituent X) lie approximately in the same spatial regions with respect to the aryl moiety: they are not hidden by any other group and appear open to possible interactions with the receptor. They are, moreover, iso-oriented with respect to the total dipole moment vector direction for the same molecules.

We have taken compound benalaxyl as a model and performed a series of rigid fits [4] matching the positions of carbonyls, central nitrogen and total dipole moment vector. For every molecule we have calculated the statistical percentage of conformations, out of the total number of energetically accepted ones, that show good superimposition of the chosen points with the corresponding ones in N° 1. Such percentages are simply the sum of population probabilities of those clusters formed by 'right' conformations. Fig. 2a shows a good relationship between biological activity and descriptive feature. In Fig. 2b

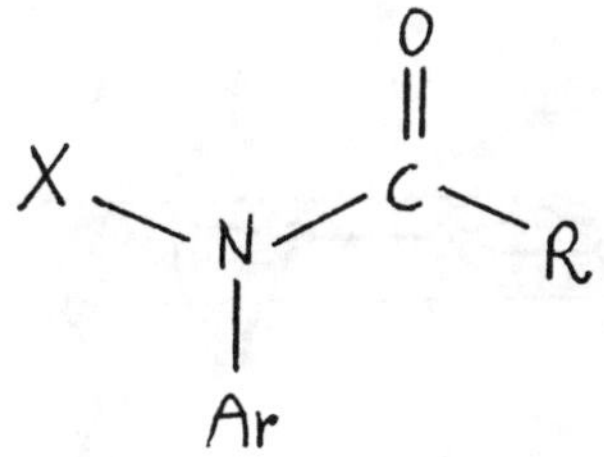

GENERAL STRUCTURE OF N-PHENYLAMIDES INCLUDING SEVERAL COMPOUNDS ACTIVE AS ANTI-OOMYCETES.

Alaninates

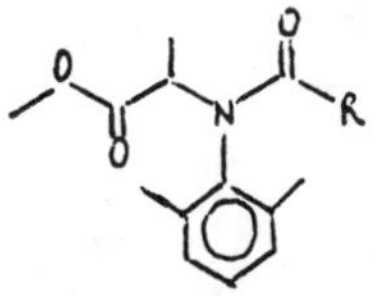

benalaxyl	(R = CH_2-Ph)
furalaxyl	(R = 2-furyl)
metalaxyl	(R = CH_2-O-CH_3)

Butyrolactones

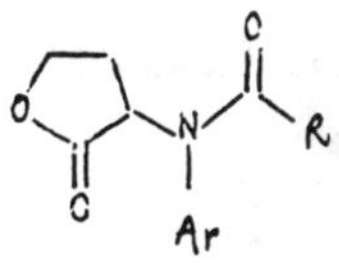

cyprofuram	(R = 9-C_3H_5 Ar = 3-Cl-C_6H_4)
ofurace	(R = CH_2-Cl Ar = 2,6-xylyl)

Oxazolidinones

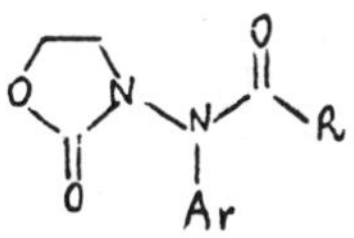

oxadyixyl	(R = CH_2-O-CH_3 Ar = 2,6-xylyl)

Fig. 1 – Considered N-PHENYLAMIDES as fungicides

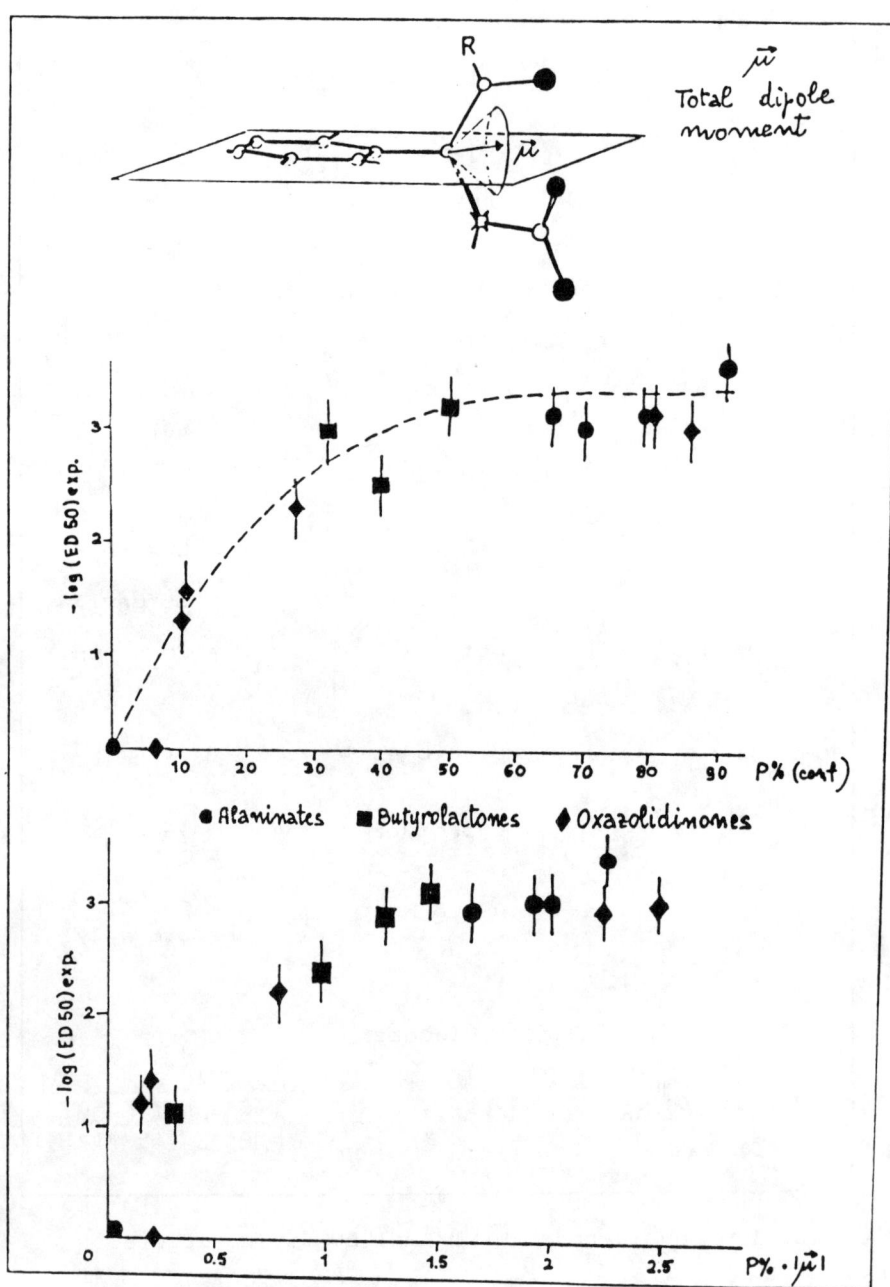

Fig. 2 - Correlations between activity and calculated descriptor

probability values have been multiplied by dipole moment vector intensity, thus improving the correlation.

In Fig. 3 the weighted average conformations in the two principal clusters for benalaxyl are shown. The weighted average conformation is a fictitious but representative conformation of the cluster corresponding to rotational angle values obtained as

$$\overline{\vartheta}_{jk} = \sum_\ell p_{j\ell}(T) \cdot \vartheta_{jk\ell}$$

where l runs over all the conformation of cluster j, k is the rotation bond index and p_{jl} is the probability of each conformation l inside the cluster j at temperature T. The dipole moment vector $\vec{\mu}$ is perpendicular to the sheet plane and is directed towards the observer's eye.

Steric differences between the two cases are well shown up by drawings.

Different types of molecular descriptors can be calculated with such a method and correlated to some peculiar property in a series of molecules. Fig. 4 gives an example of a statistical descriptor. In the case of the four considered compounds [5] it was experimentally known that thermal stability is related to their activation power for absorption of CO_2 in $H_2O-K_2CO_3$ solutions. We have found that variations in energetic levels population versus T gives an explanation for differences in stability. In Fig. 4 the energy of levels increases from bottom to top and one can see that high stability values (represented through a number of +) correspond to the rotational freedom of the molecule, described as high probability values of finding

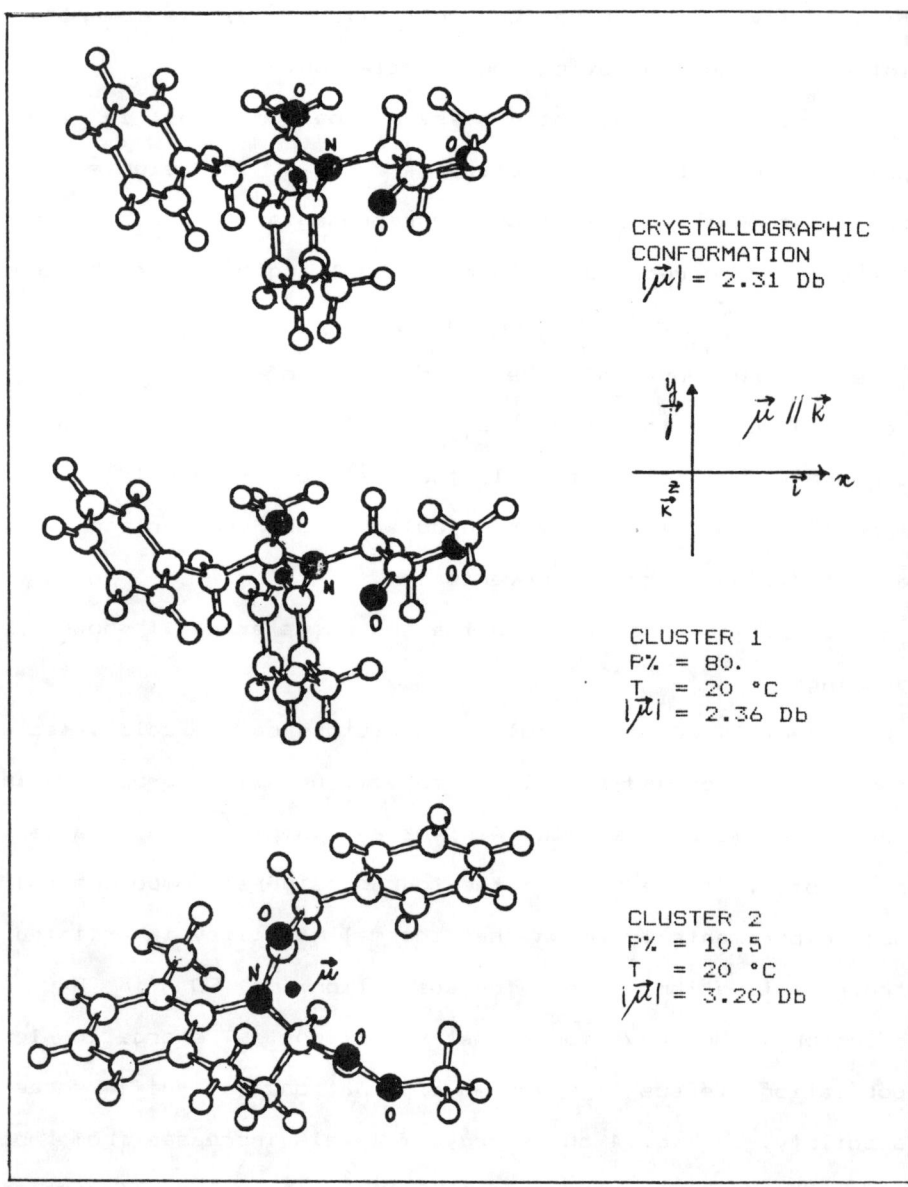

Fig. 3 - Experimental geometry and calculated conformations for the benalaxyl molecule

Compound/ Stability	Number of accepted conformations for each level (%)	Probability % of occupation for each level as function of T (°C)		
		0°C	40°C	100°C
E) OH-CH$_2$-CH$_2$-NH-COOM [+++++]	12.88 31.66 21.23 20.61 .61	0.22 3.36 13.94 76.30 .16	0.42 5.05 16.51 72.73 5.25	0.83 7.74 19.48 67.56 4.28
F) (OH-CH$_2$-CH$_2$)$_2$-NH-COOM [+++]	9.79 8.81 7.96 2.69 .61	.62 3.81 19.15 32.80 43.44	1.24 5.91 23.98 33.48 34.90	2.49 9.03 28.92 32.23 25.98
G) OH-CH$_2$-C(CH$_3$)-NH-COOH [-]	16.67 11.11 7.14 3.17 18.25	.03 .19 .88 1.98 96.90	.12 .46 1.56 2.64 95.19	.23 .87 2.34 3.26 93.16
H) HOOC-C(CH$_3$)$_2$-NH-COOH [+]	32.00 6.00 10.00 12.00 12.00	.15 .12 3.45 13.34 82.94	.42 .27 5.04 16.63 77.60	.90 .50 6.56 19.21 72.72

Fig. 4 - Statistical descriptor correlating to the thermal stability of a number of carbammates.

the molecule in high energy levels when T increases.

Other applications can be found in reference [4]. They all show the use of the above physical quantities as molecular descriptors that may characterize a molecule in a series and may be compared to corresponding quantities of another compound. The common goal is either to define in active compounds those molecular fragments where local interactions between receptor and approaching molecule presumably take place, or to determine relationships between structural parameters and biological activity. In this latter case predictions about a new compound functionality can be made starting from the values taken on by the characteristic quantities.

REFERENCES

1) G. Castellani, R. Scordamaglia, Computer & Chemistry, 8, 127 (1984)

2) A. Verloop, W. Hoogenstraaten, J. Tipker, in "Drug Design", Ed. J.A. Ariens, Vol VII, Academic Press, New York

3) F. Gozzo, L. Barino, R. Scordamaglia, paper presented to the "8th International Symposium on Systemic Fungicides and Antifungal Compounds", Reinhardsbrunn (GDR), April 27 - May 4, 1986

4) L. Barino, R. Scordamaglia, La Chimica & L'Industria, vol 68, N.11 (1986)

5) Internal Communication

Molecular Chain Flexibility and Phase Transitions in Polymers.

Luisa Barino, Raimondo Scordamaglia

Istituto G. Donegani,

Via G. Fauser 4, I-28100 Novara

Abstract

The effects linked to the transformation of thermal energy absorbed by a compound into internal rotational energy are analyzed through the CSD (Conformations Statistical Distribution) method.
The application of statistical mechanics to the allowed conformational microstates belonging to the rotational hypersurface of a molecule leads to the calculation of partition function, entropy and thermodynamic potentials of the molecular system. Conformational transitions versus temperature can then be pointed out for representative models of molecular flexible chains and correlated to phase transitions in the corresponding polymeric materials. Other physical quantities such as steric and electronic features affecting packing between molecules are studied for the models and their variations with temperature are calculated in the transition range.
Examples are given of the correspondence between calculated spectra for single polymeric chain models and experimental transition temperatures or DSC curves of some known polymers and thermotropic liquid-crystal polymers.

Introduction.

Internal rotational movements of a molecule as dependent on thermal energy absorption are analyzed with the CSD method [1].

One of the possibilities of that method consists in obtaining an evaluation of the flexibility of a molecule in vacuo. This becomes particularly interesting for molecules bearing anisotropic geometries in space, such as elongated molecules and polymeric chains. For unperturbed flexible chains it is generally assumed that the internal rotational freedom is the predominant term in molecular energy changes with temperature, i.e. total molecular energy changes are affected by thermally excited internal rotation more than by other movements of the molecule, such as roto-translations and vibrations. Another hypothesis following Flory's work [2] is that packing between chains is largely determined by conformational features in the single chain, while the effects of packing on chain conformations are less important.

The fundamental hypoteses on which the CSD approach is based can be summarized as follows:

a) The application of statistical mechanics to the allowed conformational states belonging to the rotational hypersurface of a molecule leads to the calculation of partition function, entropy and thermodynamic potentials of the molecular system, treated as a unique thermodynamic entity. The effects linked to the transformation of thermal energy absorbed by a molecule into internal rotational energy can thus be analyzed.

b) Variations of the above thermodynamic quantities versus temperature generate theoretical 'spectra' pointing out the conformational transitions that occur when there is a change in the temperature for an isolated flexible molecule.

c) Conformational transitions versus temperature for suited

representative models of molecular flexible chains can correlate to phase transitions in the corresponding polymeric materials.

d) The CSD method provides information on the allowed internal rotational motions in the molecule as dependent on the temperature, which is of interest in the study of the molecular properties of polymeric materials. Hopfinger's calculations repeatedly [3] showed the melting phenomenon in high polymers to be principally a discontinuous change in the distribution of torsion angles, accompanied by a drop in correlation between angular distributions in adjacent chains.

Representative model for flexible molecular chains.

A complete model should take into account all the processes taking place during a structural transition. These may be grouped as:

1) conformational changes or transitions within single chains;
2) cooperative processes between chains;
3) packing energy changes.

In our model only point 1 is treated explicitly, and points 2 and 3 are looked at following some appropriate assumptions on the single chain internal rotations, which simulate the reduction of chain rotational freedom due to the surroundings.

The formulated model has the following characteristics:

a) A fragment of a small number of monomers contains all the information on chain flexibility. This number of monomers should be the smallest bearing all different and numerically meaningful intramolecular interactions between nonbonded atoms,

b) Real chain length and compression effects in bulk are to some extent taken into account through a particular rotational movement strategy. Free rotatable bonds are grouped and successively rotated:

MOLECULAR FRAGMENT	ROTATED BONDS
①—②—③—④—⑤—⑥ τ_1 τ_2 τ_3 τ_4 τ_5	1) τ_1, τ_2, τ_3 2) τ_2, τ_3, τ_4 3) τ_3, τ_4, τ_5

Going from left to right, the geometry of the chain portion at rest is kept in the energy minimum conformation. Conformations from the various portions are then collected and analyzed together.

With this strategy, the chosen set of monomers is not handled as an oligomer molecule but as a representative model possessing a rotational freedom inferior to that of the oligomer and similar to that of the chain in the polymer bulk.

c) Together with thermodynamic spectra other physical quantities —such as steric and electronic features— affecting packing between molecules are calculated and their variations with temperature are analyzed in the transition ranges. In the case of liquid crystal molecules (LCM), analysis of the most probable conformations provides molecular shape changes when temperature increases and it brings out those steric features which are typical of the liquid crystal phase. In the case of polymer, the shortness of the representative fragment often cannot provide information on the geometrical shapes assumed by chains. For liquid crystal polymers (LCP) this information may be gained or lost depending on the particular compound's structure.

Molecular Descriptors.

In the following tables we have summarized all the calculated quantities i.e. thermodynamic, steric and electric. They are determined along each pathway j at every T and in thermodynamic equilibrium conditions for the molecular system. Taking the probabilities p_{ij} as weights, steric and electric quantities are calculated as weighted averages over all conformational possibilities of the fragment at each T.

THERMODYNAMIC DESCRIPTORS

1) PARTITION FUNCTION $\quad Z_j = \sum_i n_{ji} \cdot \exp\left(-\frac{1}{RT} \cdot E_i\right)$

2) CLUSTER PROBABILITY IN THE HYPERSURFACE $\quad P_j = Z_j / Z$

3) GIBBS ENTROPY $\quad S_j = -R \sum_i p_{ji} \cdot \ln p_{ji}$

4) INTERNAL ENERGY $\quad U_j = \overline{E}_j = \sum_i p_{ji} \cdot E_i$

5) HEAT CAPACITY AT CONSTANT VOLUME $\quad C_{V_j} = \left(\frac{\partial U}{\partial T}\right)_V = \left(\frac{\partial \overline{E}}{\partial T}\right)_V$

6) $\left(\frac{\partial S}{\partial T}\right) = \frac{C_{V_j}}{T}$

7) HELMHOLTZ FREE ENERGY $\quad F_j = U_j - TS_j = -RT \cdot \ln Z_j$

STERIC DESCRIPTORS

1) TORSIONS ON FREE DIHEDRAL ANGLES $\theta_{j\,min}^{(k)}$, $\theta_{j\,max}^{(k)}$
2) ANGULAR AVERAGE VALUE $\overline{\theta}_j^{(k)} = \sum_i \theta_{ji}^{(k)} \cdot p_{ji}$
3) WEIGHTED ANGULAR DEVIATIONS $\Delta\theta_j^{(k)} = \sqrt{\sum_i (\theta_{ji}^{(k)} - \overline{\theta}_j^{(k)})^2 \cdot p_{ji}^2}$
4) MOLECULAR DIMENSIONS ALONG THREE INERTIAL AXES
5) MOLECULAR VOLUME

ELECTRIC DESCRIPTORS

1) POINT CHARGE DISTRIBUTION IN THE MOST PROBABLE CLUSTERS, AS NET CHARGES ON ATOMS CALCULATED WITH CNDO/2 PROGRAM
2) TOTAL DIPOLE MOMENT INTENSITY AND DIRECTION

It is clear that when the chosen fragment is not simply a model of a polymeric chain but is a real isolated molecule, then steric and electric properties are actual descriptors of the molecular shape in space and of its packing possibilities. In the other cases generally only thermodynamic spectra and angular distribution variations are significant. A method is being developed to generate the most probable geometries assumed by an actual isolated long chain.

Examples.

In many cases we have found satisfactory relationships between calculated conformational transitions in chains and structural transitions in materials.

Some fundamental and evident considerations must be made about such a correlative approach. Theoretically calculated temperatures are expected to be lower than the experimental ones, since theory allows the molecular system to obtain its state of equilibrium at each T; this is not the case in experiments. On the contrary, the ignored intermolecular interactions between chains in the bulk do not have easily evaluable effect: on the one hand they oppose intramolecular disorder and would increase calculated transition temperatures, but on the other they corroborate the 'cascade' propagation of chain movement in bulk just as the temperature range, which starts up wide or disordered internal rotations in a single chain, is reached, thus lowering the calculated transition temperature.

The methodology cannot distinguish a-priori which type of transition can be induced in bulk by a conformational transition in the single chain. It may be a structural transition inside one phase or a transition between different phases. Steric parameters can help in selection but are not resolutive. The obtained results for a number of LCM, polymeric single chains and LCP show, however, that calculated spectra of entropy and

$\left(-CH_2-\underset{R'}{\overset{R}{\bigcirc}}-CH_2-\right)_{n=5}$			
R	R'	Tg(°C) exp./calc.	Tm(°C) exp./calc.
CN	H	90/86	270/270
Cl	Cl	110/112	380/350
CH_3	H	50-60/57	200-210/240
Cl	H	80/77	290/280
C_2H_5	H	25/25	160-170/165
H	H	80-60/60	420/380

Table I – Dependence of transition temperature on substituent groups in poly-p-xyylylene.

```
         CLUSTERS EVOLUTION PATHWAY - VARIATION OF ENTROPY
         PATHWAY NUMBER...  1

 DIFMIN  .34185650-004   DIFMAX  .29593022-003
 T(C)                    SCALE FACTOR= 10**  5          ΔS

 -60.0      29  ----------------------------*
 -40.0      20  -------------------*
 -20.0      17  ----------------*
    .0      21  --------------------*
  20.0      13  ------------*
  40.0      11  ----------*
  60.0       9  --------*
  80.0       8  -------*
 100.0      18  -----------------*
 120.0       7  ------*
 140.0       6  -----*
 160.0       7  ------*
 180.0       5  ----*
 200.0       4  ---*
 220.0       4  ---*
 240.0       4  ---*
 260.0      10  ---------*
 280.0       3  --*
 300.0       3  --*
 320.0       3  --*

         CLUSTERS EVOLUTION PATHWAY -  DELTAH-TDELTAS
         PATHWAY NUMBER...  1
                                                ΔF = ΔU - TΔS
 DIFMIN  .11565386-001   DIFMAX  .22784847-001
 T(C)                    SCALE FACTOR= 10**  3

 -60.0      17  ------------------*     \
 -40.0      21  ---------------------*  \
 -20.0      21  ---------------------*  \
    .0      16  ----------------*       \
  20.0      21  ---------------------*  \
  40.0      21  ---------------------*  \
  60.0      22  ----------------------* \
  80.0      22  ----------------------* \
 100.0      11  ----------*             \
 120.0      22  ----------------------* \
 140.0      22  ----------------------* \
 160.0      20  --------------------*   \
 180.0      22  ----------------------* \
 200.0      22  ----------------------* \
 220.0      22  ----------------------* \
 240.0      22  ----------------------* \
 260.0      13  ------------*           \
 280.0      22  ----------------------* \
 300.0      22  ----------------------* \
 320.0      22  ----------------------* \
```

Experimental values: [2]

Tg ≅ 100 °C   Tm ≅ 243 °C

Fig. 1 — Calculated spectra of entropy variation Δs and of the free energy variation $\Delta F = \Delta U - T\Delta S$ for polystyrene (iso) representative fragment $H-(CH_2-CH(\phi))_{n=5}-H$

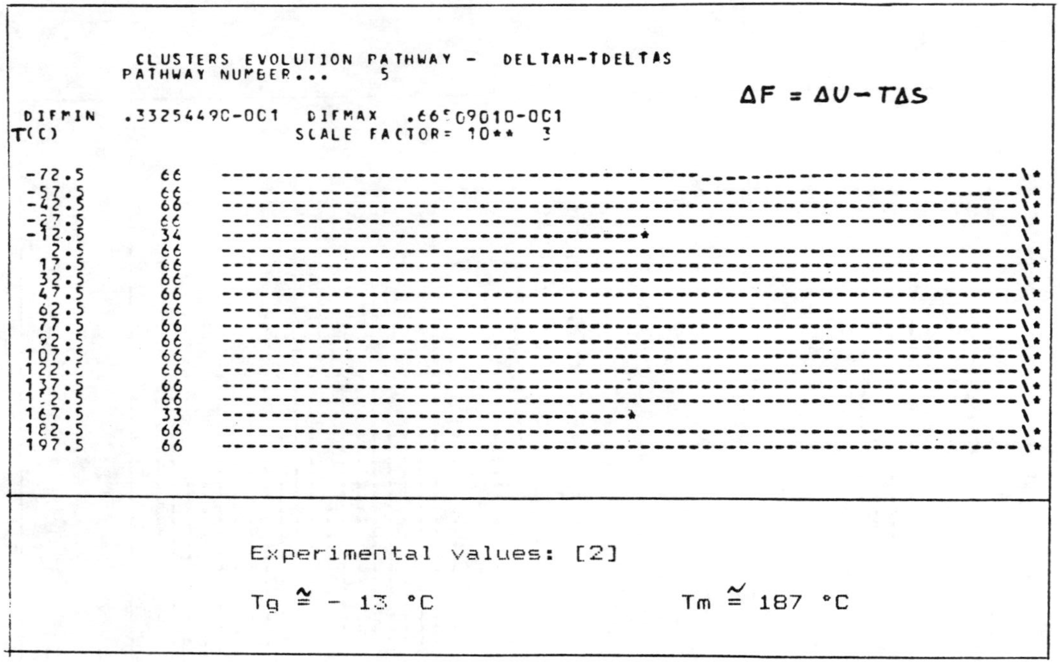
CLUSTERS EVOLUTION PATHWAY - DELTAH-TDELTAS
PATHWAY NUMBER... 5
ΔF = ΔU − TΔS
DIFMIN .33254490-001 DIFMAX .66509010-001
T(C) SCALE FACTOR= 10** 3
-72.5 66
-57.5 66
-42.5 66
-27.5 66
-12.5 34
2.5 66
17.5 66
32.5 66
47.5 66
62.5 66
77.5 66
92.5 66
107.5 66
122.5 66
137.5 66
152.5 66
167.5 33
182.5 66
197.5 66
Experimental values: [2]
Tg ≅ - 13 °C
Tm ≅ 187 °C

Fig. 2 - Calculated spectrum of free energy variation $\Delta F = \Delta U - T\Delta S$ for the polypropylene (iso) representative fragment $H\!-\!(CH_2\!-\!CH(CH_3))_{n}\!-\!H$, $n=5$

free energy variations do seem appropriate to explain structural changes and phase transitions (Tg,Tm) for the corresponding materials and, moreover, seem particularly suited to the characterization of differently substituted compounds belonging to some congeneric series: comparisons can be made between the theoretical behaviour of an experimentally known compound and of others. Consequently predictions of transition temperatures for a new compound of the series can be made.

Fig.s 1 and 2 show some calculated spectra for two well known polymers, isotactic polystyrene and polypropylene. The variation step in temperature was of 20°C for polystyrene and of 15°C for polypropylene. One can see a very satisfactory correspondence between theoretical transitions and experimental values [4].

Table 1 shows accurate research of the influence of substituent groups in poly-p-xylilene. It is an example of correlative study in an homogeneous series. The question of correspondence between conformational and phase transitions has an obvious answer and we can point out the displacements of transition temperatures due to the different substituents.

The other figures report the obtained results for a number of LCMs and LCPs and show a satisfactory compatibility of the method for the study of the latter ones. In the case of LCP, positive theoretical answers may be obtained for a molecular model formed of only two monomers or of a sort of 'dimers' of rigid portions in the chain, as already found in literature [5]. The last example shows that with a sufficient fragment length it

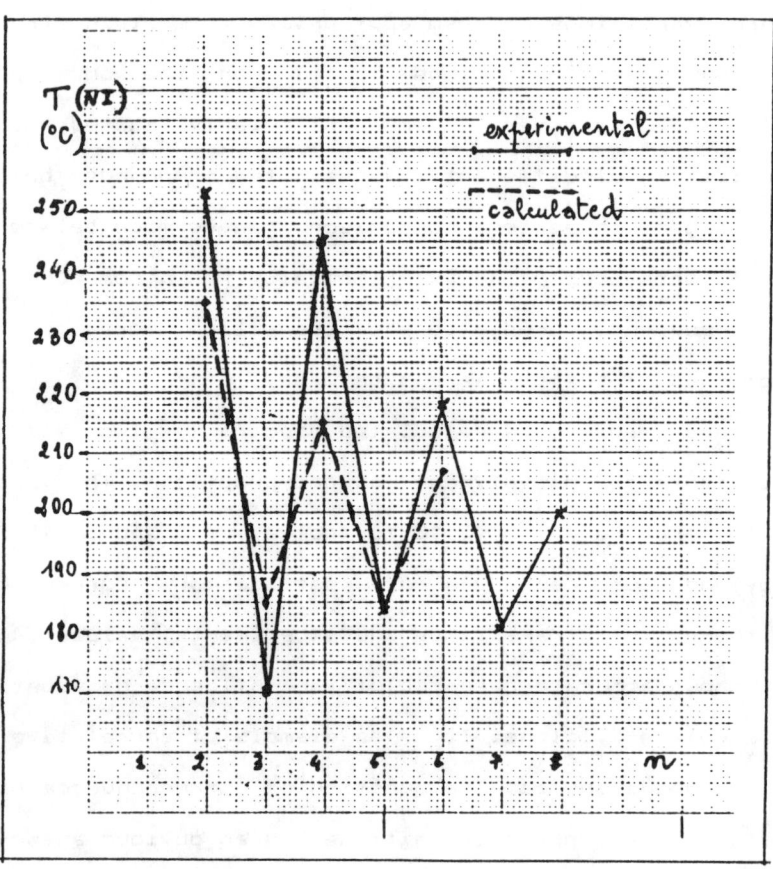

Fig. 3 — Odd-even effect of the number of bonds in the aliphatic spacer on $T_{(NI)}$ temperature (nematic-isotropic phase transition temperature) for the compound:

NC—Ø—Ø—O—(CH$_2$)$_m$—O—Ø—Ø—CN

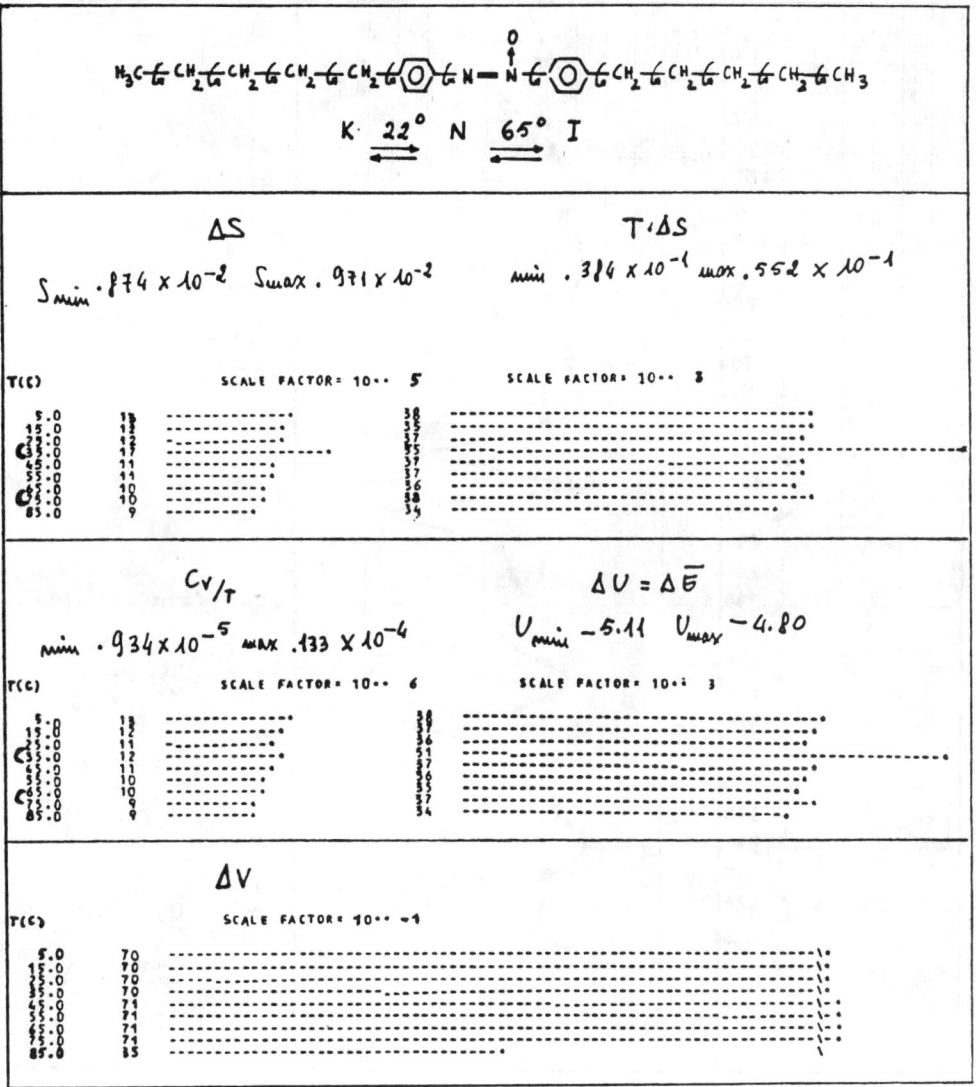

Fig. 4 - Calculated spectra for a known LCM

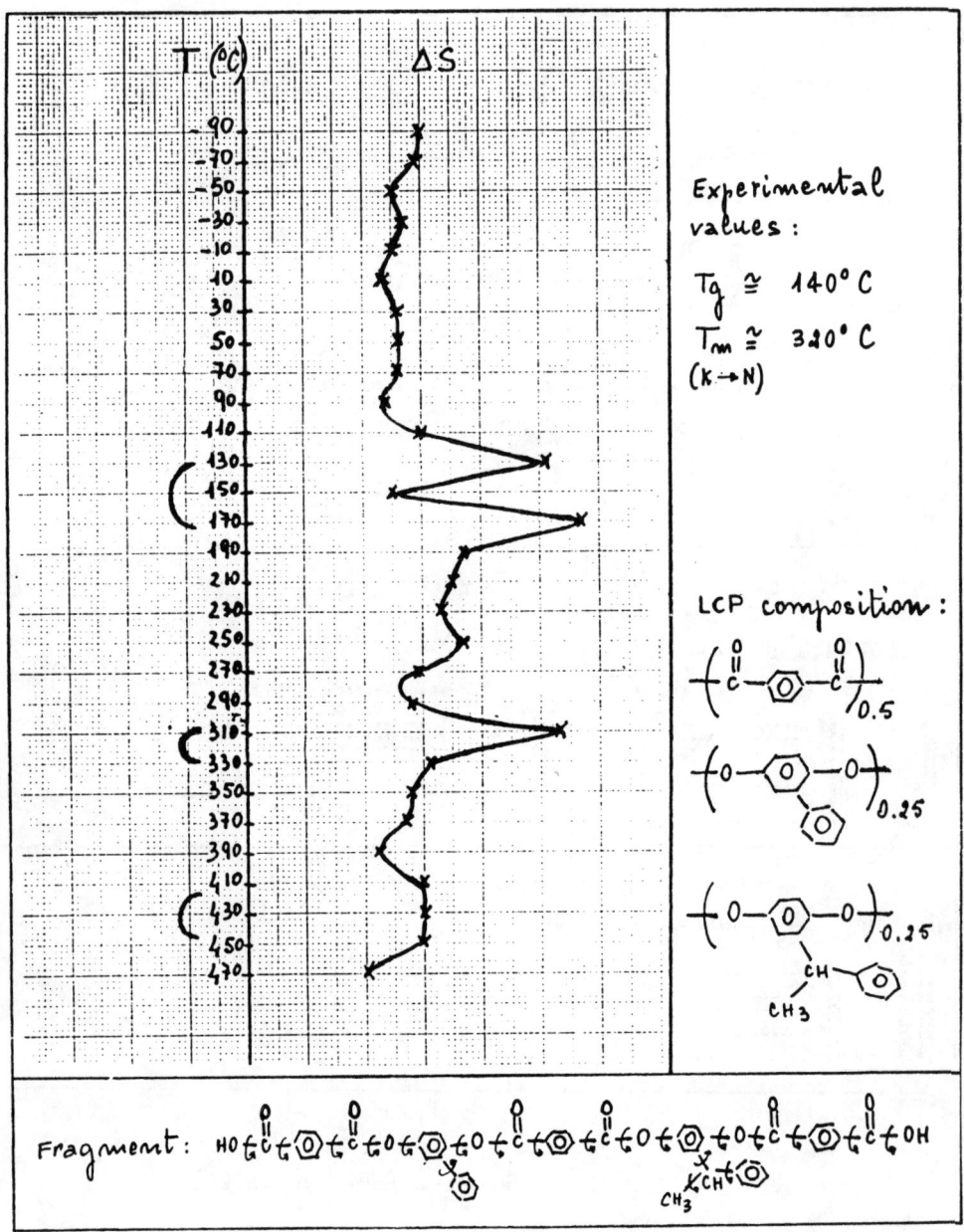

Fig. 5 — ΔS spectrum for the representative fragment indicated in the figure. Polymer is formed by three monomers in the reported % composition.

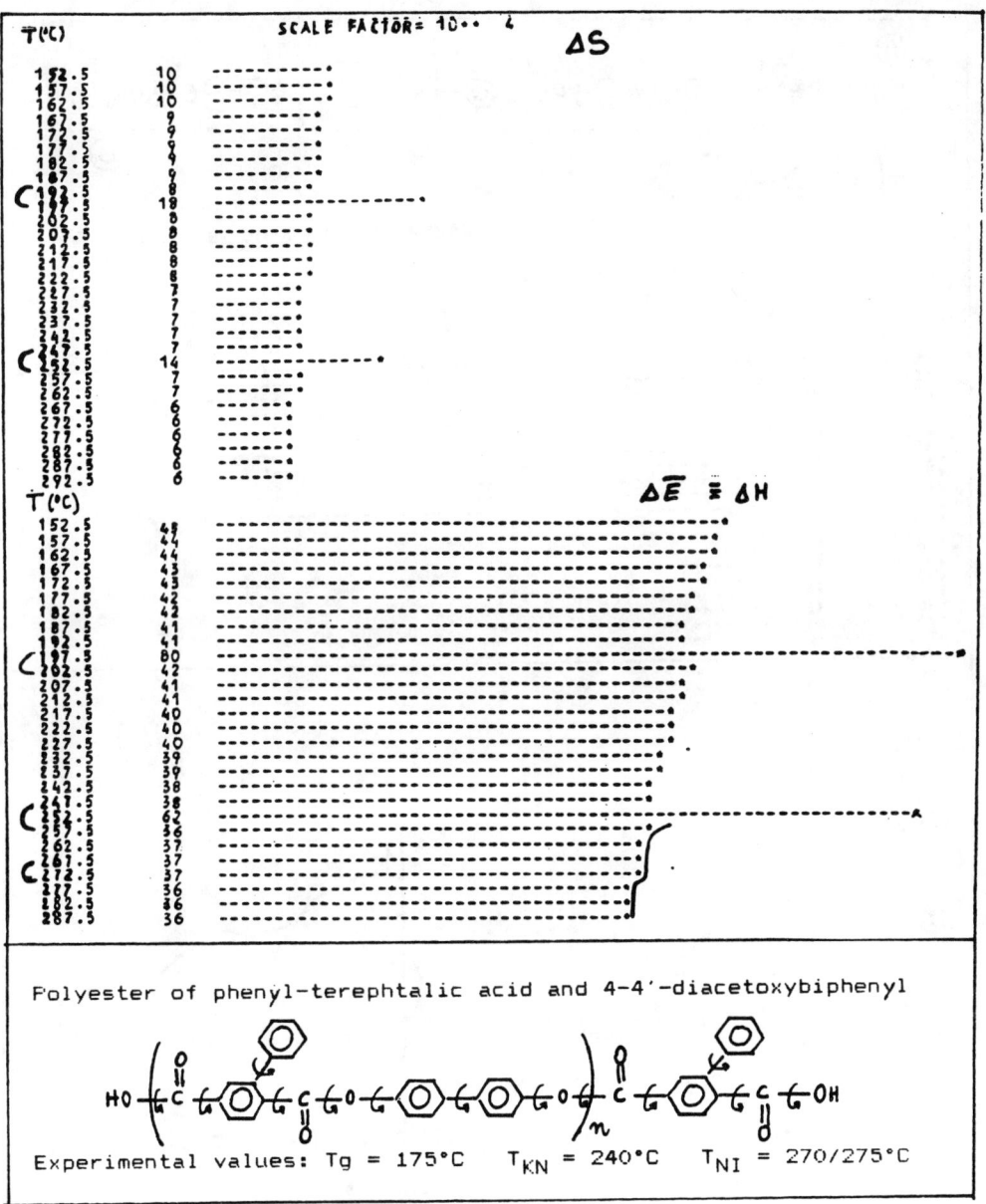

Fig. 6 - ΔS and ΔH spectra for the known indicated polyester (LCP).

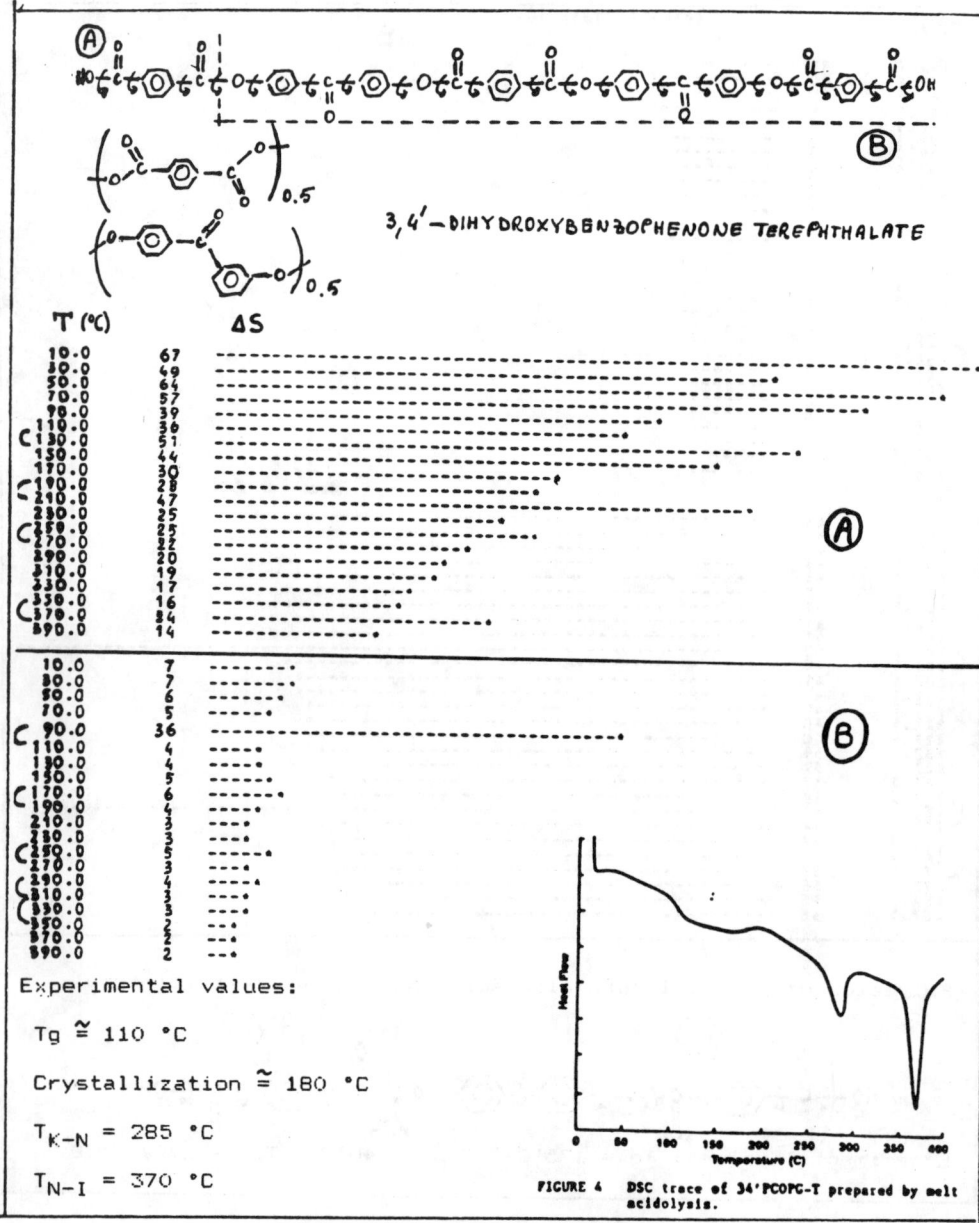

Fig. 7 - Δ S spectra for two differently composed fragment in the indicated LCP. [4]

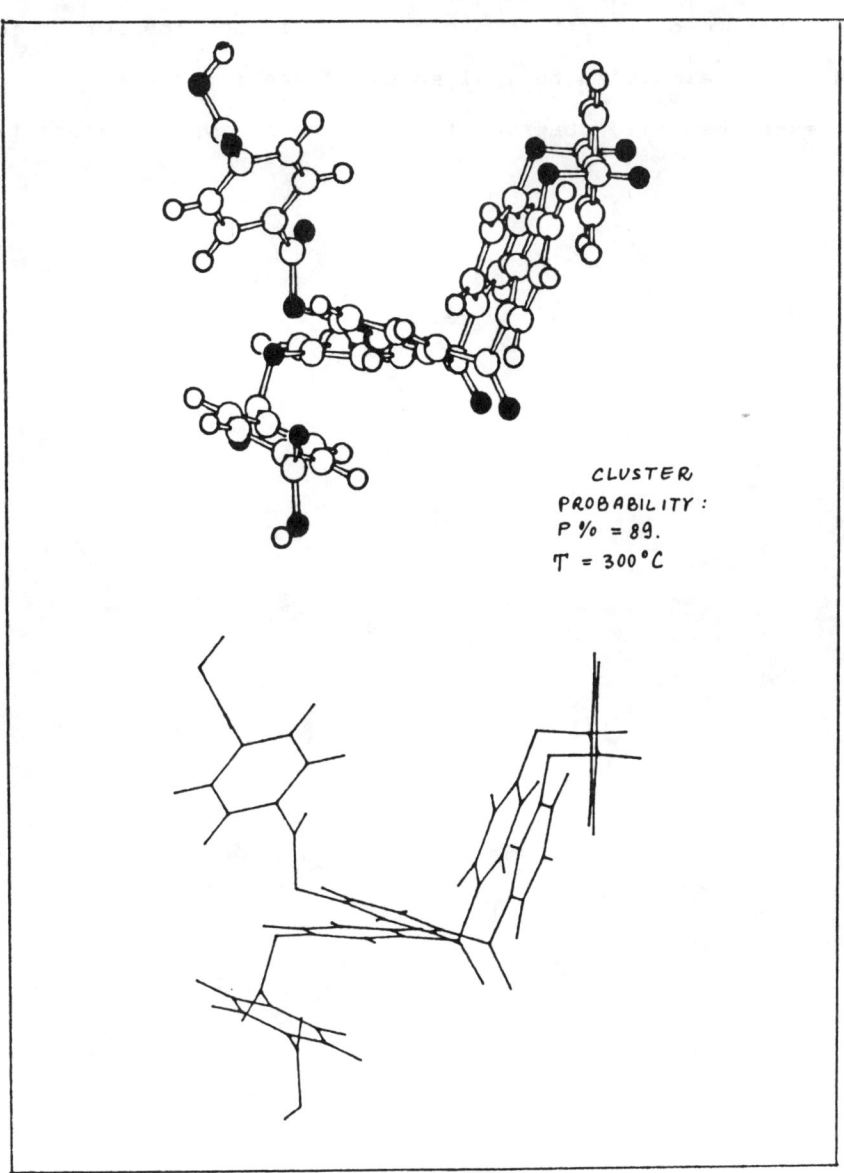

Fig. 8 — Weighted average conformation of fragment (A) of figure 7.

is possible to obtain the geometrical chain development. Fig. 8 shows the calculated helical shape of the first LCP that has been experimentally observed to bear helix-type conformations [6].

References

1) R. Scordamaglia, L. Barino, "Statistical Distribution of Molecular Conformations and Its Applications in QSAR Research", this issue.

2) P. J. Flory, "Statistical Mechanics of Chain Molecules", Wiley-Interscience, New York, (1969)

3) A. J. Hopfinger, "Conformational Properties of Macromolecules", Academic Press, New York (1973);
A. J. Hopfinger, R. A Pearlstein, P. L. Taylor and F. P. Boyle, J. Macromol. Sci. - Phys. B 26(3), pp. 359-386 (1987)

4) ATHAS (Advanced Thermal Analysis - A Laboratory for Research and Instruction), Fourth Report (1987)

5) J. A. Buglione, A. Roviello and A. Sirigu, Mol. Cryst. Liq. Cryst., Vol. 106, pp 169-185 (1984)

6) K. H. Gardner, C. R. Gochanour, R. S. Irwin, W. Sweeny and M. Weinberg (E. I. du Pont de Nemours & Co.), "Conformational and configurational characteristics of 3,4 - DIHYDROXYBENZOPHENONE TEREPHTHALATE", Proceedings of the International Conference on Liquid Crystal Polymers, Bordeaux, July 1987.

GENERAL ASPECTS OF COMPUTER-AIDED SYNTHESIS PLANNING

Rainer Moll

VEB Chemiekombinat Bitterfeld, GDR-4400 Bitterfeld,
Research and Development, FO/S

Abstract

"Computer-Aided Synthesis Planning" (CASP) is a special and important branch of computer chemistry. At present there are different theoretical and methodological approaches for solving the problems connected with the development of practicable programs. CASP-systems can be knowledge-, logic- or rule-oriented, and they can work with a backward- (retrosynthetic) or forward-strategy. A further difference is, whether they operate in batch- or dialogue-mode.

In this paper the fundamental principles and some advantages and disadvantages are discussed. Moreover, the "RDS"-program, developed at the Academy of Sciences of GDR and tested, improved and used in VEB CKB, is described in some detail.

At present we are in a phase of very rapid development of microelectronics and computer techniques and their influence in all fields of social life. Especially in working processes, not only rationalization and increased efficiency can be achieved, but entirely new possibilities also emerge. This is important for organization and management, for controlling of production processes and, to an increasing extent, also for science and technology.

In the course of this development chemistry has played a leading role. On the one hand methods of data processing were used rather early, on the other new fields of application can be seen almost daily. In the case of chemical research, not only the collection, storing and processing of data are important, but also the direct aiding of creative events, for instance by means of so-called expert systems and methods of artificial intelligence. Both terms are controversial, but they are in common use and we are not here to discuss whether they are correct or misconceptions.

In Fig. 1 an overview is given of the most important applications of computers in chemical research and development. The capability of the computer to store and to process huge amounts of data in a reproducible and "unprejudiced" manner comes from the fact that information and documentation is one of the oldest and more familiar fields of application, e.g. as literature storages and retrieval systems (online-systems), compound-databases, structure information-databases or spectral-databases. Furthermore the acquisition, storage and processing of measured values led to the development of new techniques of process controlling, laboratory automation, structure elucidation and other fields of modern analytical chemistry. In this context a new scientific discipline was established, chemometrics [1]. The object of this is to apply mathematical and statistical methods to chemical problems, to develop optimized techniques and to gain a maximum of information from analytical and other data, e.g. by means of classification methods, like pattern recognition. Other fields, scientific calculations or quantum chemistry, also play an important role. Finally, after modest and rather esoteric first steps, a new branch arose, comprising simulation of chemical reactions, modelling procedures with the aid of theoretical considerations, techniques for structure representation and so on, i.e. methods to aid creative and decision-making processes.

The differences between A and B on the one hand, and C and D on the other, rest, simply speaking, on the fact that the former complexes serve the storing and processing data, whilst the latter produce data. It is a matter of course that the fields are interacting and cannot be seen isolated from one another. The areas, which are summed up under letter D and which are shown in Fig. 2 in greater detail, are at present in very rapid development. Within the framework of the structure elucidation of chemical compounds, the simulation of spectra and their comparison with the data of known substances are fields of increasing importance [2]. Mass spectroscopy seems the most advanced in this respect, but efficient programs also exist for vibrational, NMR- and ESR-spectroscopy. In the area of drug design or the development of pesticides, the relations between molecular structure and biological activity are not sufficiently known. For this reason, mathematical-statistical methods have been developed, the so-called "Quantitative Structure-Activity Relationships" (QSAR), with the aid of which the correlation between biological effects and some structural properties can be modelled [3]. Another field of simulation is metabolite prediction, e.g. for a pharmacon in the human body or for a pesticide in the environment; but these methods are only at their beginning. The main goal of all these efforts is to come to molecules with desired and predictable properties, i.e. a "Molecular Design".

The mentioned complexes have gained additional attractivity, but also additional methodological possibilities, from molecular graphics. Currently many kinds of programs are on the market, not only for mainframes but also, in increasing measure and quality, for PC's. The development of animation programs was a great success because with these methods movement in all directions and

Figure 1

APPLICATION FIELDS OF COMPUTERS IN CHEMICAL RESEARCH

A: INFORMATION AND DOCUMENTATION

- literature storages / retrieval systems (e.g. online systems)
- compound databases
- structure information databases
- spectral databases

B: ACQUISITION AND PROCESSING OF MEASURED VALUES (CHEMOMETRICS, TECHNOMETRICS)

- process control
- lab automation
- structure elucidation

C: SCIENTIFIC-TECHNICAL CALCULATIONS / QUANTUM CHEMISTRY

D: AIDING OF CREATIVE AND PROBLEM-SOLVING PROCESSES

- simulation procedures
- modelling procedures
- structure representation

Figure 2

SIMULATION / MODELLING / DESIGN

- simulation of chemical reactions / synthesis planning

- spectra simulation

- quantitative structure-activity relationships (QSAR)

- metabolites prediction

- molecular design

- structure representation / molecule manipulation and animation

---

the manipulation of structures became possible. This can be of special interest for the representation of receptor sites and effector-receptor interactions, e.g. by "mapping" and "docking" and the modelling of molecules (cf. [4]).

If one reviews all the mentioned and unmentioned fields of application of computers in chemical research, it seems justified to speak of a new special branch, "Computer Chemistry", which can be defined [5] as:

- Acquisition, description, representation, utilization, and producing of chemical knowledge by means of data and information processes using proper methods and conceptions.

An area in which this very definition comes true in a fairly strong sense is Computer-Assisted Synthesis Planning. Before its different methods and concepts are considered, some general remarks have to be made. The preparation of new compounds or the more efficient synthesis of known substances are the fundamental goals of synthetic chemistry. An important motive is the obtaining of compounds with useful properties. The expenditure of time, material costs, and intellect is considerable. Efforts have been and are being made to lower the expenditure and to rationalize the research activity. One possibility is synthesis planning. Neither the term, nor the application are new; what are new are the methods and the technological prerequisites. It was the merit of COREY to see how synthesis planning can be done more effectively, more systematically and more successfully [6]. He profited by the great advances in the computer technology at the end of the Sixties, i.e. hardware came onto the market with sufficient computational speed, and also from developments in the software field. Furthermore, great experience in the strategic planning of the synthesis of rather complex molecules, especially natural products, had been accumulated, not least by COREY himself.

At this point it is necessary to say something about the information-theoretical aspects of the topic. From the point of view of an information scientist, chemical synthesis offers a series of suitable prerequisites for the solution of its problems with the aid of data and information processing:

1. The problem space is sufficiently great and complex.

2. Its elements can be classified and formalized.

3. The problem can be analyzed logically.

4. Chemistry has its own language with standardized rules (formula language).

5. The problems can be treated with algorithms.

However, the great, maybe excessive expectations could not be fulfilled until the early Eighties. Only in this decade was a true breakthrough achieved. The reasons are the following:

1. Complexity larger than previously assumed.

2. Lack of information.

The problem of complexity can be demonstrated by a comparison between the game of chess and synthetic chemistry, published a few years ago by GASTEIGER [7], which is modified and supplemented by another very complex system, the weather forecast (for details see Fig. 3). It can be seen that synthesis (and weather forecast too) are more complex than chess, though this is complex enough. The last line of the second column in Fig. 3 leads directly to the next problem, the lack of information.

For a successful planning of a synthesis two types of information are needed:

- structural information, which is sufficiently available in general, and

- reactivity information.

The latter is not available to such an extent and so accurately as demanded for the purposes of synthetic planning. The reason for this is that there is no general valid, consistent and complete quantitative formularizable theory of chemical reactivity.

But it would be fully unjustified to conclude that synthesis planning would not be capable of giving useful problem solutions. On the one hand, one can overcome or evade the difficulties mentioned with suitable conceptual approaches, and this has been done rather successfully. On the other hand, a decisive impetus is coming from synthesis planning for the further development of the theory of chemical reactivity.

Before the program package "RDS", used in the CKB, is explained, a short overview about other programs or interesting systems shall be given. Since there is rather extensive literature on these matters (see [8] and [9] for reviews), there is no need to go into great detail. Generally speaking, the fundamental differences between the systems consist in the way they try to solve the problems of reactivity and evaluation. The criteria of interest in this context are summarized in Fig. 4.

Knowledge-oriented systems are those programs which operate on the basis of learning from literature, either in the form of reaction libraries or in the form of special databases, such as transform-libraries and so on. On the other hand we have formal- or logic-oriented systems which function by means of pure logical or mathematical operations, without any direct reference to chemical knowledge. That means, in the first case one can use chemical experience as a whole, with all its facts, but also with all its limitations; in the second, if one follows its propagators,

Figure 3

COMPARISON BETWEEN CHESS GAME, SYNTHESIS PLANNING AND WEATHER FORECAST

	CHESS	SYNPLAN	WEATHER FORECAST
SPACE	64 fields	multidim. energy surfaces	atmosphere
OBJECTS	chessmen	atoms	measured values
ARRANGEMENTS	positions	molecules	inhomogeneities
MOVEMENTS	moves (number very great, but limited)	reactions (number practically unlimited)	formation and degration of inhomogeneities
RULES	well defined	heuristic/ empirical	heuristic

Figure 4

CHARACTERISTICS OF PROGRAMS FOR SYNTHESIS PLANNING

knowledge-   formal (logic)-   rule-oriented

    with evaluation   without evaluation

    batch   interactive (dialogue)

synthesis planning   reaction prediction
(retrosynthetic;   (synthetic;
backward-strategy)   forward-strategy)

                 bi-directional

                 bi-lateral

one is able to simulate not only the conceived, but also the conceivable reactions and comes in this way to new solutions. A compromise between both sides will be accomplished by the third type, called rule-oriented systems. These systems use chemical knowledge in a formal and abstract manner and with special rules.

A further problem is whether or not one sets in the program evaluation steps, e.g. as options, heuristic rules, special restrictions, etc. Working without any evaluation, the computer can produce much chemical nonsense, which has to be eliminated by the user. In the other case, it is very difficult to find the right type of evaluation and it is then possible that reasonable solutions are suppressed by the program itself. A further point is whether the system works in a batch- or dialogue-mode; in the latter case the chemist has the possibility of influencing directly the simulation process, but also with its prejudices. Moreover, there are differences in the programs, whether they make, strictly speaking, synthesis planning (working in retrosynthetic direction, backward strategy), or reaction prediction (synthetic direction, forward strategy). In the first route a target molecule is the starting point of the simulation and the system produces precursor- or "son"-structures, which can serve themselves as starting molecules for the next step and so on. In this manner so-called synthesis trees are built up (Fig. 5). In the other mode the reaction is simulated and the possible products are predicted, if special compounds are input as starting materials. A new principle, first formulated by UGI [10], works bilaterally, i. e. forward- and backward-strategy work simultaneously. After the input of a target and possible or conceivable starting substances, two synthesis trees are built up in both directions and form a network; on its branches the chemist can find the shortest and the most efficient way for the solution of his synthetic problem.

The first running, and at present maybe the most used system is LHASA (Logic and Heuristic Applied to Synthesis Analysis), developed by COREY and coworkers [11]. It works by means of so called "transforms", which represent a special kind of reaction description and which are derived from literature; they are the knowledge base of the program. For the evaluation many heuristic and strategic rules are used; the mode is interactive and the direction retrosynthetical. WIPKE and his group developed SECS (Simulation and Evaluation of Chemical Synthesis) [12], which resembles LHASA in its general approach. The main differences consist in the type of transform-library and in the set of evaluation rules. For both programs input languages have been written: "CHEMical TRANslator" for LHASA and "ALCHEM" for SECS; the output is in both cases graphical. It has to be noted that SECS is used and further developed under the designation CASP (Computer-Assisted Synthesis Planning) by some chemical and pharmaceutical companies in FRG and Switzerland. Of special interest is SYNCHEM/2 [13], with a large synthesis library (in form of "synthems") and strong restrictions as evaluation; it works in the batch-mode forward and backward. It is interesting to know that its inventor GELERNTER is not a

Figure 5

SYNTHESIS TREE (RETROSYNTHESIS)

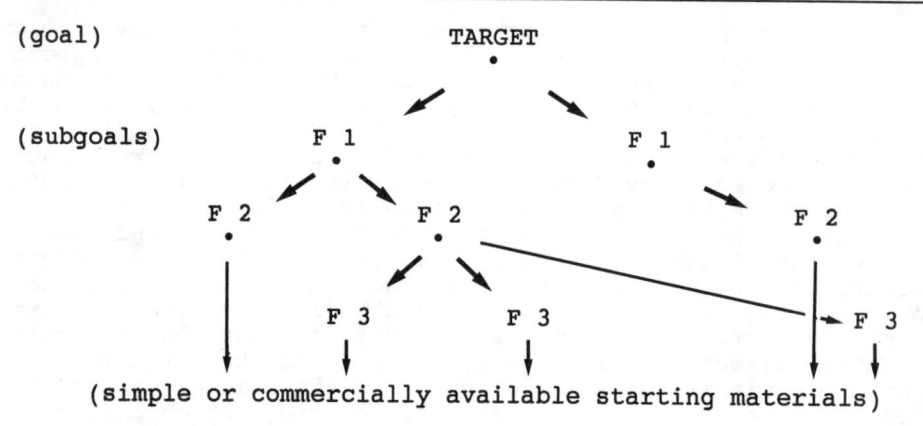

Figure 6

RDS (REACTION DESIGN SOFTWARE) - GENERAL PRINCIPLE

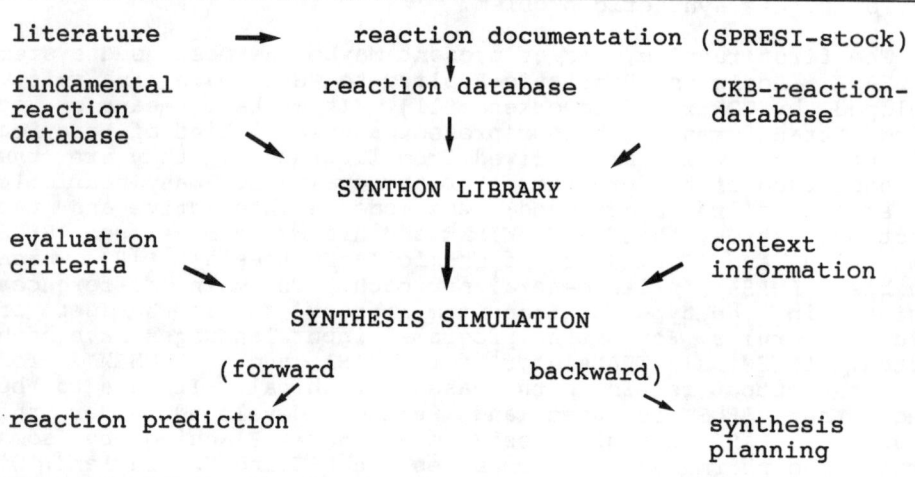

chemist, but an information scientist. A special position is held by DARC (Description, Acquisition, Retrieval and Computer-aided design) by DUBOIS [14]. It is not a "pure" synthesis planning program, but it is thought of as a complete system for the modelling of chemical reactions. A fairly interesting package, which is extended and developed permanently, is JORGENSEN's CAMEO (Computer-Aided Mechanistic Evaluation of Organic Reactions) [15]. The system is built up strictly from mechanistic points of view, has its own individual modules for each reaction type, and operates forward exclusively.

A completely different approach on the basis of the model of "constitutional chemistry" was initiated by UGI and is further developed by the Munich groups. The programs EROS and IGOR and their predecessors are formal- or logic-oriented, work by means of so-called BE- and R-matrices and without chemical knowledge. Because the general principles are described in literature rather extensively [16] there is no need to give an overview of the formalisms here. Whilst EROS (Elaboration of Reactions of Organic Synthesis), written by GASTEIGER and coworkers [17], now operates with a system of physico-chemical parameters for evaluations, IGOR (Interactive Generation of Organic Reactions) from UGI's group [18] runs without any evaluation. EROS works in the batch-mode, a dialogue version is also available, and retro- and forward-synthetic. IGOR is completely interactive and runs in both directions too. It has to be mentioned that in UGI's group a new system - RAIN (Reaction And Intermediate Network) [10] - is under way, working in the bilateral approach with the PMCD (Principle of the Minimal Chemical Distance) [19] for formal evaluation.

RDS (Reaction Design Software), developed by WEISE at the Academy of Sciences of GDR and further improved in the VEB Chemiekombinat Bitterfeld, is a rule-oriented system, i.e. a compromise between the knowledge- and the formal-oriented approach. The structure and the mode of action of the program-suite is depicted in Fig. 6. The knowledge base is a reaction documentation, abstracted from literature in the form of SPRESI-stock (SPRESI is the German acronym for "Storage and Retrieval of Structural Chemical Information) [20]. By means of a program written for this task, from this documentation a reaction database is generated which only contains reaction equations (with a minimal set of bibliographical data). This database again serves as the source for the "synthon library", which is also derived from special programs. In the same way a CKB-reaction-database (containing unpublished results from the CKB research) and a fundamental reaction database (containing mainly "textbook"-chemistry) are source for the synthons. It has to be stated, that the databases represent the knowledge base and the synthon library the formal one. The simulation or planning process is made with the synthons whereas it is possihle to work without evaluation or with a hierarchic system of rules. The evaluation complex consists of mechanistic and heuristic rules and is tested permanently and is in further development. Moreover, so-called context information is being developed (e.g. about other reactive sites in the mole-

Figure 7

SYNTHON PRINCIPLE

- REACTION EQUATION:

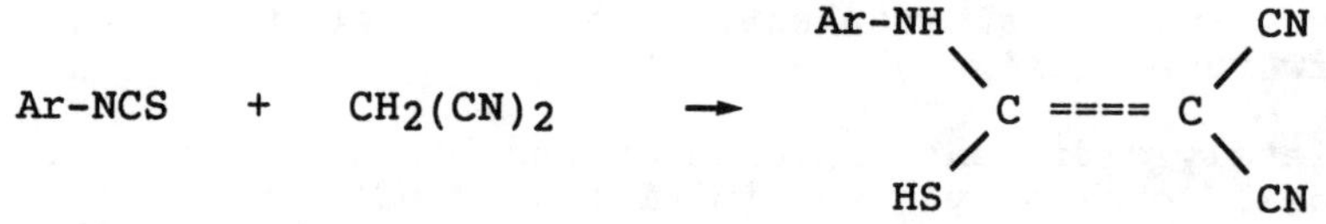

$$Ar\text{-}NCS + CH_2(CN)_2 \longrightarrow \underset{HS}{\overset{Ar\text{-}NH}{>}}C = C\underset{CN}{\overset{CN}{<}}$$

- SYNTHON EQUATION:

$$\text{-}NCS + \text{-}CH_2\text{-} \longrightarrow \underset{\text{-}S}{\overset{\text{-}NH}{>}}C = C<$$

Figure 8

EXAMPLES OF AHMOS-SIMULATIONS

Example 1:

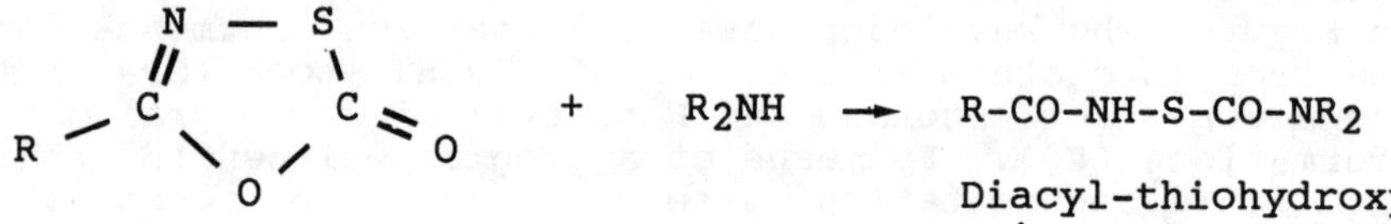

$$\underset{R}{\overset{N=S}{>}}C\underset{O}{<}\underset{O}{>}C=O \; + \; R_2NH \; \longrightarrow \; R\text{-}CO\text{-}NH\text{-}S\text{-}CO\text{-}NR_2$$

Diacyl-thiohydroxyl-amines

Example 2:

$$R\text{-}CO\text{-}CH=CCl_2 + Ph\text{-}NH\text{-}NH_2 \longrightarrow R\text{-}CO\text{-}NHNHPh + HCl + [HC \equiv CCl]$$

Carboxylic acid hydrazides

Figure 9

CARSA (COMPUTER-ASSISTED RESEARCH IN SYNTHESIS AND APPLICATION

WIFODATA	substance- and finding-documentation storage for active agents QSAR-program-library storage for LFE-parameters etc.
RDS	synthesis planning
Retrieval Systems	e.g. for patent-abstracts etc.
Application Software	mathematical-statistical calculations molecular orbital caculations graphic tools programs for in-house databases etc.

cule). The RDS-package is designed for the forward- and backward-strategy; a bilateral version is intended to be developed in the near future.

The central idea of RDS, the synthon approach, is explained in Fig. 7. The term, originally coined by COREY and differently used in literature, means, in RDS, those parts of a molecule which are directly involved in the reaction and not identical on both sides of the reaction equation. A synthon equation is derivable from each reaction equation. In the simulation process according to the "General Synthon Substitution Principle" the synthons are copied one upon another by means of a "similarity decision procedure". By this method precursor- or product-structures and synthesis trees are generated.

The RDS program-suite has already rendered its capability for solution of problems. It is plain that examples from current research work cannot be given here, but two problems, which were performed with the predecessor system called AHMOS (Fig. 8), can be shown. In the first example the formation of a new type of compounds, the diacyl-thiohydroxyl-amines, was predicted and confirmed by synthesis [21], and in the second case a surprising finding, the formation of carboxylic acid hydrazides from dichloro-vinyl-ketones and phenylhydrazine, was recognized [22].

It is a matter of course that in the VEB CKB, synthesis planning is not separated from other fields of computer application in research and development. The RDS-programs are a component of a total system, called CARSA (Computer-Assisted Research in Synthesis and Application) (cf. Fig. 9). This is a modular assembled, but integrated CHEMO-CAD-system, which is extended and further developed permanently, and at present consists of the following constituents [5]:

- WIFODATA (a database and program library)

- RDS (synthesis planning)

- Retrieval systems

- Application software packages.

An extension to methods of molecular modelling is planned. The first three components are implemented on CKB's mainframe, the other programs operate on PC's.

To conclude, it can be stated that there is no doubt that computer-aided synthesis planning, like other methods in computer chemistry, will play an important role in the future. The potential consists in the intellectual help for the chemist and the capability to support him in creative and problem-solving processes.

List of References

[1] M.A. SHARAF, D.L. ILLMAN and B.R. KOWALSKI, "Chemometrics", J. Wiley, New York (1986).

[2] B. ADLER, "Computerchemie", VEB Deutscher Verlag fur Grundstoffindustrie, Leipzig (1986).

[3] R. FRANKE, "Theoretical Drug Design", Elsevier Sci. Publ. B.V., Amsterdam (1983).

[4] a) C. TOSI, R. SCORDAMAGLIA, L. BARINO, G. RANGHINO, R. FUSCO and L. CACCIANOTTI, Chim. Ind. (Milan) **69** (1-2), 68 (1987); b) T. Gund and P. GUND, Mol. Struct. Energ., 319 (1987).

[5] R. MOLL, P. KEMTER, U. LINDNER, D. SCHOENFELDER and A. WEISE, Chem. Techn. (Leipzig), in press.

[6] E.J. COREY, Quart. Rev. **25**, 455 (1971).

[7] J. GASTEIGER, Chim. Ind. (Milan) **64**, 714 (1982).

[8] J.H. WINTER, "Chemische Syntheseplanung", Springer Verlag, Heidelberg (1982).

[9] C. TROMBINI, Chim. Ind. (Milan) **69** (5), 82 (1987), and literature cited therein.

[10] E. FONTAIN and I. UGI, Chem. Letters, 37 (1987).

[11] E.J. COREY, A.K. LONG and S.D. RUBENSTEIN, Science **228**, 408 (1985).

[12] P. GUND, E.J.J. GRABOWSKI, D.R. HOFF, G.M. SMITH, J.D. ANDOSE, J.B. RHODES and W.T. WIPKE, J. Chem. Inf. Comp. Sci. **20**, 88 (1980).

[13] H. GELERNTER, S.S. BHAGWAT, D.L. LARSEN and G.A. MILLER, in "Computer Applications in Chemistry" (S.R. Heller and R. Potenzone, Eds.), p. 35f, Elsevier Sci. Publ. B.V., Amsterdam (1983).

[14] J.E. DUBOIS, C. MERCIER and A. PANAYE, Acta Pharm. Yugosl. **36**, 135 (1986).

[15] A.J. GUSHURST and W.L. JORGENSEN, J. Org. Chem. 51, 3513 (1986), and literature cited therein.

[16] I. UGI, J. BAUER, J. BRANDT, J. FRIEDRICH, J. GASTEIGER, C. JOCHUM and W. SCHUBERT, Angew. Chemie **91**, 99 (1979).

[17] J. GASTEIGER, M.G. HUTCHINGS, B. CHRISTOPH, L. GANN, C. HILLER, P. LOEW, M. MARSILI, H. SALLER and K. YUKI, Topics Curr. Chem. **137**, 19 (1987).

[18] J. BAUER, R. HERGES, E. FONTAIN and I. UGI, Chimia **39**, 43 (1985).

[19] C. JOCHUM, J. GASTEIGER and I. UGI, Angew. Chemie **92**, 503 (1980).

[20] A. WEISE and H.G. SCHARNOW, Z. Chemie **19**, 49 (1979).

[21] G. WESTPHAL, A. KLEBSCH, A. WEISE, U. STERNBERG and A. OTTO, Z. Chemie **17**, 295 (1977).

[22] R. MOLL and A. WEISE, unpublished results.

DETECTION AND STRUCTURAL DESCRIPTION OF THE DEEPEST MINIMA IN A POTENTIAL ENERGY HYPERSURFACE

R. Fusco, L. Caccianotti and C. Tosi

Istituto G. Donegani, Via Fauser 4, I-28100 Novara

Abstract

A detailed description of the algorithms used in an improved version of the LECSA program (Low-Energy Conformational Space Analysis) is presented. In particular the potentialities of this methodology in describing both structural and physico-chemical features as functions of thermodynamic variables are pointed out.

1. Introduction

A combined method of random sampling and local quasi-Newton minimization for seeking low energy minima on the energy hypersurface of a molecule was described in a previous paper [1]. It aimed at solving one of the main problems of conformational analysis, i.e. the determination of the most stable molecular conformations. This method, called LECSA (Low-Energy Conformational Space Analysis), is complementary to another method, developed over the last few years in our laboratory, aimed at detecting the global minimum of intramolecular potential energy functions [2]. They have been applied to a number of molecules, among which ethyl methyl phosphate [1], a di-deoxyribose mono-phosphate fragment [3], the analgesic hepta-peptide dermorphin [4], the antitumour drug methotrexate [5].

The joined application of both methods has proved to be one of the most powerful tools that are available today for the study of molecular conformations [6]. Because of the rather short time elapsed since their first development, the potentialities of these two methods have not been fully exploited as yet, and we are currently engaged in the effort to increase their performance.

We present here an improved version of the LECSA algorithm which gives some more information about the deepest minima of potential energy surfaces. Our exposition consists in three parts: in the first we discuss in some detail the concept of stability as referred to molecules; in the second we describe an implementation of the method presented in [1] for the localization of low-energy minima; in the third we propose an algorithm that allows us to explore the surrounding of each minimum in order to evaluate the statistical weight of the associated region and to characterize it from both the geometric and the physico-chemical points of view.

2. On the concept of molecular stability

In statistical thermodynamics, the concept of stability can be replaced by the concept of "high probability". According to the Boltzmann distribution law, the probability p of finding a molecule in a state with energy E at the absolute temperature T is

$$p = e^{-(E/kT)}/Q \qquad (1)$$

where $Q = \int_\Omega e^{-(E/kT)} dq^N$ is the partition function for the isolated molecule, k is the Boltzmann constant and N is the number of the degrees of freedom of the molecule, indicated by q. For a molecule containing n atoms, N is equal to 3n-6, and the conformational energy can be regarded as a function $E(l,\theta,\phi)$ of the internal coordinates, where l are the (n-1) bond lengths, θ are the (n-2) valence angles and ϕ are the (n-3) internal rotation angles. In general the energies involved in deformation of bond lengths and valence angles are one order of magnitude larger than for rotations about single bonds. Therefore, to a first approximation, the dimensions of the conformational space can be decreased by holding the former kind of geometric parameters fixed: the conformational energy can then be considered as a function of the latter only, $E = E(\Phi)$. Eq. (1) tells us that, at any given value of the absolute temperature T, the lower is the energy of a state, the higher the probability of finding the molecule in such a state will be. In other words the lowest-energy states are the most populated ones and the problem of finding the most stable conformations limits itself to localization and characterization of the low-energy regions of the potential energy hypersurface $E(\Phi)$.

For each value of the temperature of the system, we can assume that all the conformations with energy higher than a threshold E_t give a negligible contribution to the partition function. The validity of this hypothesis and, consequently, the choice of E_t, essentially depends on the degree of approximation used in the calculation of the partition function on the one hand, and the value of T on the other.

The hyperplane $E = E_t$ intersects the energy function $E(\Phi)$ in many points, which divide the conformational space in sub-domains. Let us denote as Ω_L (where L stands for low) the ensemble of the sub-domains for which the conformational energy is lower than E_t, and Ω_H (with H standing for high) those where the energy is higher than E_t. Therefore the partition function can be written in the following way:

$$Q = \int_{\Omega L} e^{-[E(\phi)/kT]} d\phi + \int_{\Omega H} e^{-[E(\phi)/kT]} d\phi$$

If the second integral is regarded as negligible, we can calculate an approximate value Q_α of Q:

$$Q_\alpha = \int_{\Omega L} e^{-[E(\phi)/kT]} d\phi$$

thereby making an error $\delta Q = Q - Q_\alpha = \int_{\Omega H} e^{-[E(\phi)/kT]} d\phi$. The advantage of calculating the approximate value Q_α of the partition function consists in the possibility of reducing the size of the conformational space to be explored, by focusing our attention only on that part of the energy hypersurface which lies under the energy-threshold hyperplane. How can the error in the calculation of the partition function be evaluated without exploring the Ω_H subspace? An upper limit δQ_{u1} of this error is given by:

$$\delta Q_{u1} = \int_{\Omega H} e^{-(E_t/kT)} d\phi = V_{\Omega H} e^{-(E_t/kT)}$$

where $V_{\Omega H}$ is the volume of the Ω_H subspace. But in general:

$$\forall \phi \in \Omega_H \quad E(\phi) > E_t$$

and consequently:

$$e^{-[E(\phi)/kT]} \ll e^{-(E_t/kT)}$$

wherefrom:

$$\delta Q = \int_{\Omega H} e^{-[E(\Phi)/kT]} d\Phi \quad \ll V_{\Omega H} \; e^{-(E_t/kT)} = \delta Q_{u1}$$

We will see later how it is possible to evaluate δQ_{u1} in order to be sure that, for a given E_t, its value is small enough. Although the choice of E_t affects also the size and the shape of the sub-domains, for a prefixed threshold it is possible to evaluate the statistical weight of each low-energy region. This statistical weight represents the probability p_i of finding the molecule in the i-th low energy region (L_i):

$$p_i = \int_{\Omega L i} e^{-[E(\Phi)/kT]} d\Phi \; /Q \qquad (2)$$

In this way it is possible to learn how the molecule is arranged in its own conformational space.

3. Detection of low-energy minima

We show here an improved version of the algorithm described in [1], based on the statistical analysis of the angular distribution of low-energy minima. Starting from an arbitrary initial conformation, a random number generator is used for producing other conformations each of which is, potentially, the starting point for an energy minimization. We have used the word "potentially" because the minimization procedure is only carried out if either of the two following conditions occurs:

i) $E(\Phi) - \epsilon_g < E_t$

where ϵ_g is the value of the lowest minimum already detected and E_t is a prefixed energy threshold, or

ii) $|\Phi - \beta_i|^2 > r_i^2 \quad \forall \, i$

where β_i is the centre of mass of the maxima (m^-_{ij}, m^+_{ij}) along the N Φ_j-axes nearest to the minimum μ_i (Fig. 1):

$$\beta_i = (1/2) \, \Sigma_j \, (m^-_{ij} + m^+_{ij})$$

and r_i is the average value of distance between μ_i and these

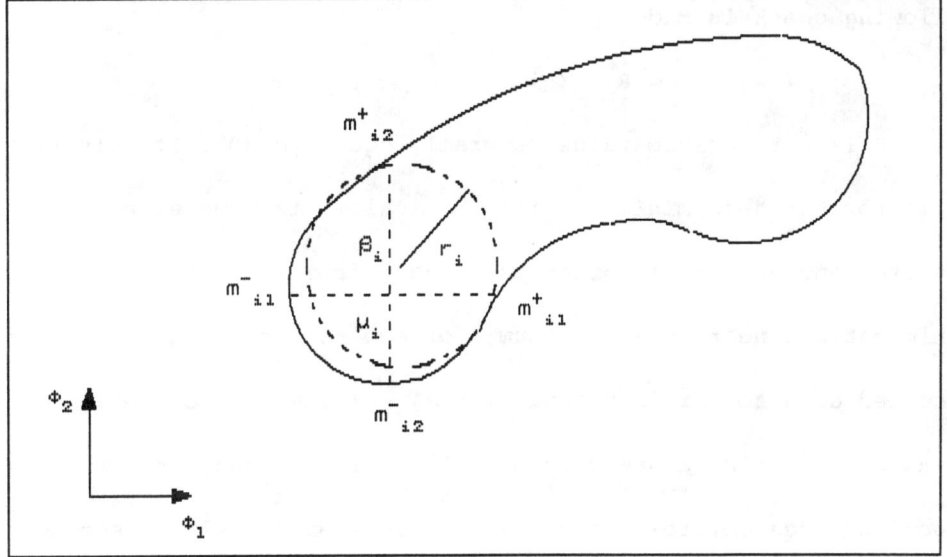

Fig. 1 : The line of the maxima nearest to the minimum μ_i intersects the axes in the four points m^+_{i1}, m^-_{i2}, m^-_{i1}, m^+_{i2}. β_i is the center of mass of these points and r_i is the radius of the forbidden hypersphere. All the starting conformations within this sphere will be skipped.

maxima

$$r_i^2 = (1/2N) \Sigma_j [(m^-_{ij} - \mu_{ij})^2 + (m^+_{ij} - \mu_{ij})^2]$$

In the event of failure of these two tests, a counter is increased and, if its value is lower than a prefixed number, a new random conformation is generated and tested. Otherwise, through an energy minimization, a minimum μ is detected. In order to test whether this minimum has yet been localized, its distance from all the minima previously detected is evaluated and the following check is made:

$$|\mu - \mu_i|^2 < R^2 \quad \forall \ i$$

Here R is a tolerance value generally equal to 30°. If this test fails for the j-th minimum, its r_j value is increased by a prefixed amount δr in order to prevent from performing further minimizations near this minimum; otherwise the minimum μ is accepted as a new minimum, and if $E(\mu) < \epsilon_g$, ϵ_g is updated.

At this point a new random conformation is generated and the algorithm goes on, following the procedure previously described. When a prefixed number M of minima has been detected, a statistical analysis of angular distribution of the minima is performed. From now on, the random conformations are generated only in the populated angular ranges. In this way it is possible to skip high-energy regions, and the search for new minima will be carried out only in those parts of conformational space where the probability of finding them is higher. A schematic representation of this methodology is described in Fig. 2 for a bidimensional case. Such angular ranges are updated through a new statistical analysis whenever the number of the minima detected is a multiple of M. The exploration ends when the method cannot generate further valid starting points.

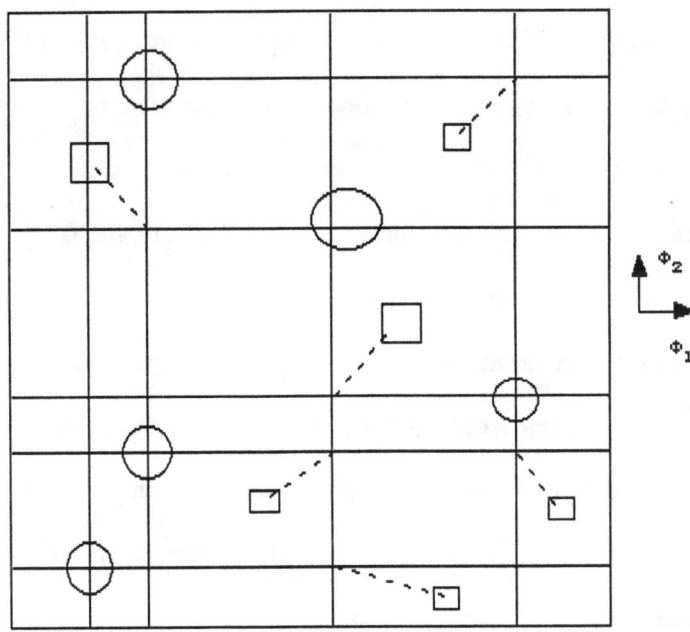

Fig 2. : Grid of potential starting points built by the intersections of the line defined by the angular values of the already detected minima (circles); further minimizations for localizing new minima (squares) can start only from the meshpoints.

At this point a test on the exhaustiveness of the exploration is carried out. If the molecule includes n rotating groups j, each one containing a K-fold symmetry axis, each minimum must show a K-order degeneration, where

$$K = \Sigma_{j=1,n} K_j$$

If this test is satisfied, the exploration is considered completed.

4. Partition function and statistical weights determination

Once N low-energy minima have been obtained, the problem of the determination of their statistical weights arises. It consists in calculating the value of the integral shown in Eq. (2) through a numerical method for each minimum whose energy is lower than E_t, taking the value ϵ_g of the global minimum as the zero of the energy scale. The integration domains are bordered by the hyperlines, given by the intersection of $E = E_t$ hyperplane with $E = E(\Phi)$, which surround all low-energy minima. Each domain can contain one or more minima.

We propose here a method for discretizing such domains by performing a random walk inside each of them. As shown in Fig. 3, the algorithm starts from the minimum μ_i and begins to sample the points by proceeding along a random direction by regular steps δ.

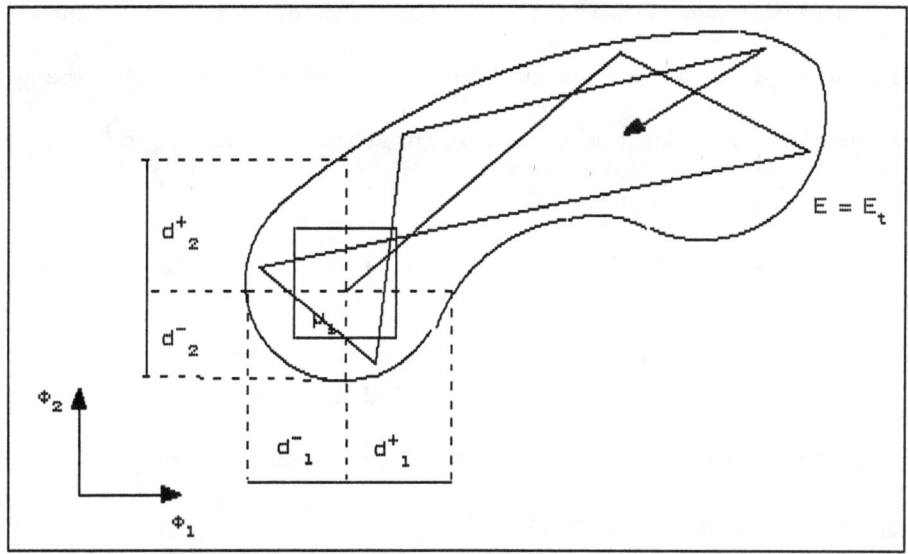

Fig. 3 : Random walk inside a potential well bordered by the hyperline $E = E_t$.

Whenever an intersection with the energy-threshold hyperline occurs, a new random direction inside the region to be explored is generated and the sampling goes on. The random walk stops when a prefixed value ρ of the density of points is exceeded. The control on the density value is performed by checking that the following condition occurs:

$$N/V \geq \rho$$

where N is the number of sample points which fall within a region of known volume V inside the integration domain. The volume V is calculated as the product of the half sums of the distances of the considered minimum from the intersection points of the axes j with the hyperline $E = E_t$

$$V = (1/2) \Pi_{i=1,N} (d^+_i + d^-_i)$$

as shown in Fig. 3.

According to the concepts expressed in the Introduction, the probability p_j of finding the molecule in the j-th low energy region can be calculated as

$$p_j = (1/Q) \Sigma_{k=1,Mj} e^{-(E_k/kT)}$$

where

$$Q = \Sigma_{l=1,N} \Sigma_{m=1,Ml} e^{-(E_m/kT)}$$

Here N is the total number of the low-energy regions, M_j (M_l) is the number of sample points within L_j (L_l). The upper limit of the error evaluation of the partition function $\delta Q_{u1}/Q$ can be evaluated as:

$$\frac{\delta Q_{u1}}{Q} = \frac{(n-v) e^{-(E_t/kT)}}{\Sigma_{i=1,v} e^{-(E_i/kT)} + (n-v) e^{-(E_t/kT)}}$$

where n is the total number of points that would give the exact value of Q for a certain density value, whereas v are the points actually used in the calculation.

5. Some applicative examples

In order to illustrate some of the potentialities of these algorithms we present a number of examples coming from their

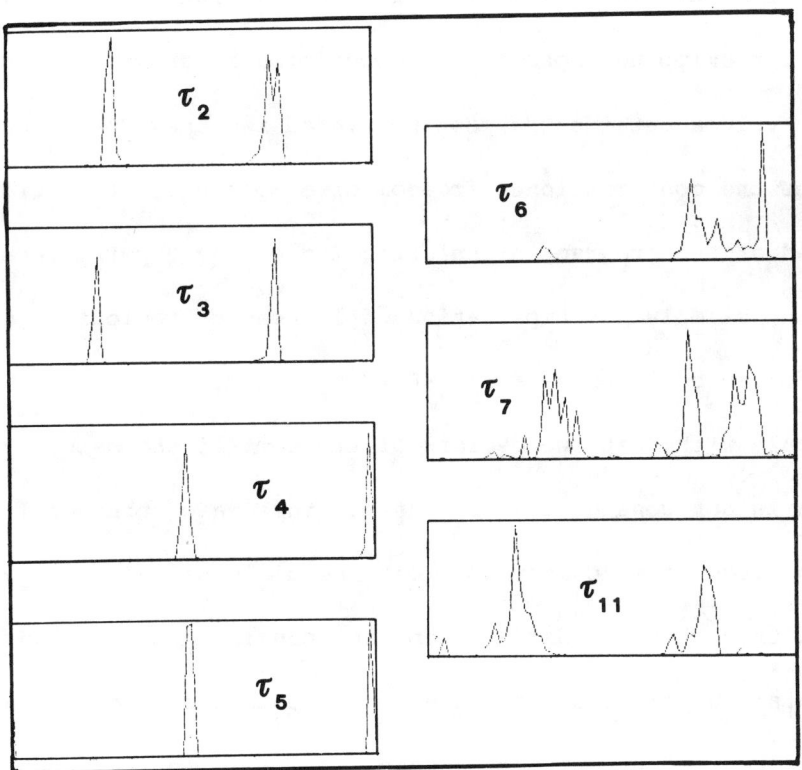

Fig. 4 : Angular frequency distribution for internal rotation angles of the methotrexate molecule [5].

applications to some conformational analysis problems. Fig. 4 shows the angular frequency distribution of the minima relative to some internal rotation angles of the methotrexate molecule [5]. It is readily apparent that for some internal rotation angles (τ_2, τ_3, τ_4, τ_5) there are a few narrow regions highly populated, while for others (τ_6, τ_7, τ_{11}) the range of accessible angular regions is shallow. This information is used by the program for skipping those starting conformations which include angular values outside the most populated regions. Consequently, the lower the conformational freedom of a molecule, the greater the speed of the program in exploring its energy hypersurface.

As an example of exploration of low-energy regions we chose the conformational analysis of EMP (Ethyl Methyl Phosphate) [1], a molecule with just two degrees of freedom (if the methyl group rotation is not considered), because in this case the analogies emerging from the comparison of the bidimensional energy map obtained through a standard systematic scansion method and the representation of the selected sample points of the low-energy regions localized by our methodology are readily apparent and impressive. Fig. 5 shows the contour map on the left and the sample points obtained by a random-walk exploration for an E_t

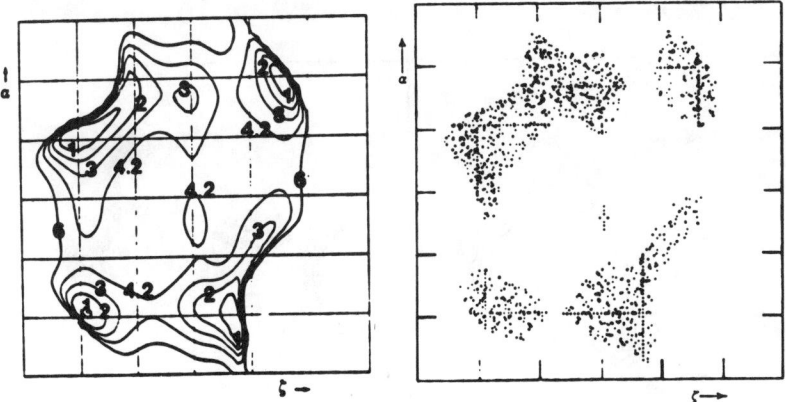

Fig. 5 : Comparison between the energy map (a) for EMP molecule and the map of the sample points used by the random-walk algorithm for the exploration of the low-energy regions (b) [1].

---

value of 4 Kcal/mol on the right.

For a number of degrees of freedom bigger than two, this kind of graphical representation cannot be obtained. However, it is always possible to select all the conformations belonging to a particular low-energy region and to show each conformation by its 3-D molecular representation. By superimposing all these conformations, through a least-squares fitting method, we can get an idea of how the molecular structure changes for different E_t values. Fig 6 shows the methotrexate molecule in the global minimum potential well for two values of the energy threshold.

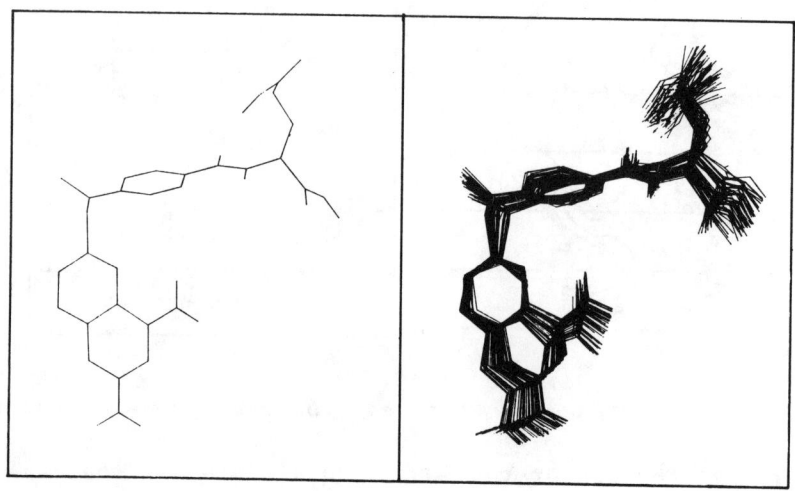

Fig. 6 : The methotrexate molecule in the global minimum potential well: a) conformation obtained for $E_t = 0$ Kcal/mol; b) superimposition of the sampling conformations corresponding to $E_t = 2$ Kcal/mol.

Bibliography

[1] R. Fusco, L. Caccianotti and C. Tosi: Nuovo Cim. $\underline{8}$ D, 211 (1986).

[2] C. Tosi, R. Pavani, R. Fusco, F. Aluffi-Pentini, V. Parisi and F. Zirilli: Rend. Accad. Naz. Lincei, Classe Sci. Fis. Mat. Nat., Ser. VIII, $\underline{78}$, 149 (1985).

[3] C. Tosi, R. Fusco and L. Caccianotti: Nuovo Cim. $\underline{8}$ D, 219 (1986).

[4] G. Ranghino, C. Tosi, L. Barino, R. Scordamaglia and R. Fusco: J. Mol. Struct. Theochem $\underline{164}$, 153 (1988).

[5] C. Tosi, R. Fusco and L. Caccianotti: J. Mol. Struct. Theochem (1988), in press.

[6] C. Tosi, R. Fusco, L. Caccianotti and V. Parisi: J. Comp. Chem., to be submitted.

MONTE CARLO SIMULATION OF THE SOLVATION OF A RIBONUCLEOTIDE

Graziella Ranghino

Istituto G. Donegani, Via Fauser 4, I-28100 Novara

The present paper describes the application of the Monte Carlo method in statistical mechanics and particularly the use of this simulation tool for the calculation of mean values of observables in the study of liquid state and of solution properties.

The basic theory states that if we generate a large number of microstates of an ensemble of particles (atoms, molecules), such that each j-th state neither depends on the (j-1)-th state, nor on the time of the observation, we define "Markov chain" this sequence of states, and it can be proven that the mean value <F> of the observable F is equal to the sum over the whole chain of the f(j) values multiplied by the probability of the state when the chain length goes to infinity [1]:

$$<F> = \Sigma_{j \to \infty} f_j \, p_j$$

$$p_j = [\exp(-\beta U_j)] / \{\Sigma_j [\exp(-\beta U_j)]\}$$

p_j = probability of the j-th state

f_j = value of F in the j-th state

U_j = energy of the j-th state

The energy U is computed by an analytical function including the van der Waals and Coulombic terms describing intermolecular interaction. The problems to face in the simulation are the usual problems of large-scale calculations: CPU time and mass memory on one side; reliable potentials and well-tailored physical model to describe the real conditions, i.e. number of molecules, volume, number of configurations, degrees of freedom, on the other. In order to give an example of the computational procedure and of the type of problems that can be tackled by Monte Carlo, we briefly report on a work which is still in progress on the structure of the solvent in the region of the anticodon loop of the t-RNA(asp). This work is done in collaboration with Dino Moras' group at the Institut de Biologie Moléculaire et Cellulaire in Strasbourg, who determined the crystallographic structure of the nucleic acid [2], and with Georges Wipff and colleagues at the Institut Le Bel of the same University.

The biological problem which is at the origin of this study is connected with the transfer of the genetic code from the DNA to the ribosomes in the protein biosynthesis.

In this flow of information, the t-RNA anticodon carrying the code for the correct amino acid is bound to the ribosome and must recognize its complementary codon on the m-RNA: this mechanism is very efficient, since the probability of error is very low. One hypothesis is that the assembly of the two nucleic acids is by no means a single step process, but instead the codon of m-RNA forms Watson-Crick and wobble base pairs with the anticodon, and in this way the recognition of the genetic triplet goes via a minihelix [3].

In the crystal of the t-RNA two independent molecules have intermolecular contact at the level of the anticodons due to the crystallographic symmetry.

This region is particularly rich in solvent and it is disordered: the Monte Carlo simulation will give us information on the mean solvent structure of this portion of the molecule which could be used in the X-ray refinement.

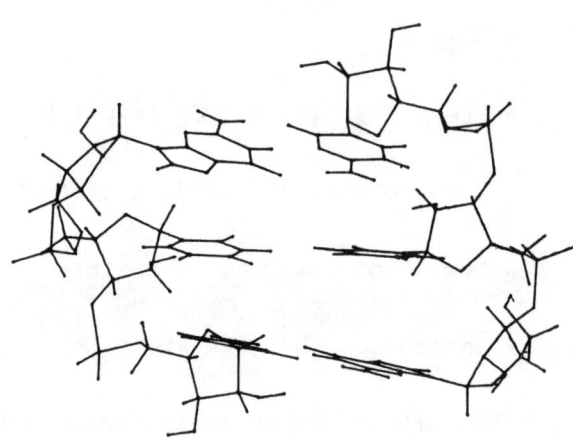

Fig. 1. The hexanucleotide $(C-U-G)_2$ as resulting from X-ray data.

Moreover, by means of simulation we hope to understand if and how the solvent is involved in the intermolecular interaction and thus in the recognition, since the crystallographic arrangement perfectly mimics the t-RNA - m-RNA interaction.

The physical model we chose is described in Fig. 1. This is just a little bit of the large macromolecule, which has 76 bases arranged in the well-known cloverleaf structure [4].

We add hydrogen atoms to guanine-34, uridine-35, cytidine-36 and we notice that the second triplet generated by the symmetry operation forms Watson-Crick base pairs C-G and G-C, but at the level of U-U there is no hydrogen bond, instead the two hydrogens point one toward the other. We went through an energy refinement of this particular base pair and we ended up with the structure displayed in Fig. 2, in which G-C and C-G have the same arrangement and U-U forms one hydrogen bond. Since the conformational change is small and the optimization is done without solvent, we performed all the Monte Carlo simulation with the original X-ray structure and 250 water molecules.

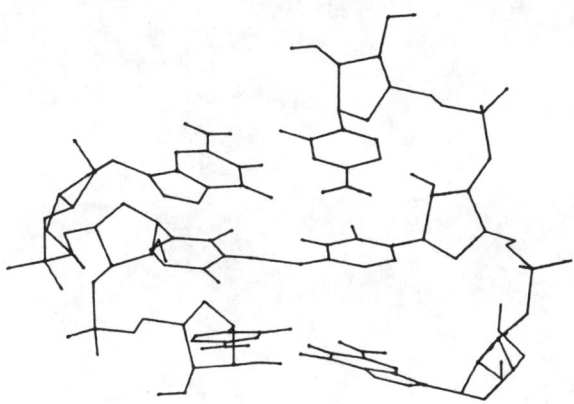

Fig. 2. Structure of the hexanucleotide after energy refinement of the U-U base pair.

The interaction potential for solute-water and water-water interaction was taken from Clementi's work [5,6]. We performed $2.5 * 10^6$ moves, and we did not include counter ions at this stage since we did not have any clear evidence of the position of the cations which are present in the crystal and we could not account for them correctly in terms of interaction potentials.

The cluster of 250 molecules appears to be sufficient to describe the first and second solvation shell of the nucleotide and the water-water energy has a reasonable value for such a negatively charged solute.

At this stage the analysis of the data is not completed, but we can anticipate some qualitative results: first that there is no fixed position for water between the two helices, but only outside, on the sugar-phosphate-sugar backbone. There is a clear network of hydrogen bonds between water molecules, as has already been observed both by X-ray and by computations [7,8] in the case of nucleotides.

By numerical integration over the whole volume of the probability of having an oxygen atom on a tight 3-D grid lattice, we assign to each node of the grid a density of atoms per $Å^3$ and at the end we can select a number of highly probable oxygen positions.

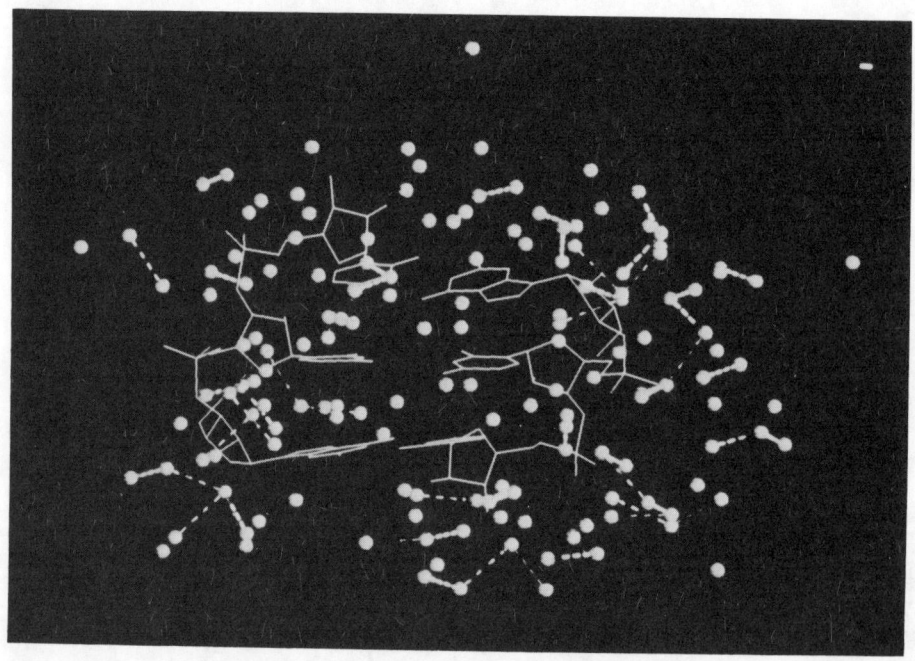

Fig. 3. Average picture of the solvent molecules in the first and second solvation shell.

The total number of occupied nodes is ca. 150: as seen in Fig. 3, some of the solvent molecules which are in the first and second solvation shell are hydrogen-bonded to each other and/or to the solute; a few of them appear to be too close to the solute and too tightly bound to it. These positions on the surface of the nucleotide are good candidates for ions and therefore they should be compared with the X-ray data of highest occupancies for solvent molecules. For the other 100 molecules there are no high-density positions, and therefore they are not included in this picture of the first and second shell of solvation.

As mentioned, work is still in progress and therefore we lack details on solvation shells and energy values, but the data we already have allow a comparison with the experimental data on the solvent structure and a detailed analysis of the role of the solvent in the pairing of the trinucleotides.

Acknowledgements

The author would like to thank Dino Moras, Philippe Dumas, Jean-Claude Thierry and Eric Westhof for making available the X-ray coordinates, for suggestions and discussions, and Jean-Marie Wurtz for assistance at the computer graphics.

References

[1] Metropolis N., Rosenbluth A.W., Rosenbluth M.N., Teller A. and Teller E., J. Chem. Phys. **21**, 1087 (1953).

[2] Moras D., Comarmond M.B., Fischer J., Weiss R., Thierry J.C., Ebel J.P. and Giegé R., Nature **288**, 669 (1980).

[3] Alden C.J. and Arnott S., Biochem. Biophys. Res. Commun. **53**, 806 (1973).

[4] Saenger W., "Principles of Nucleic Acids Structure", Springer-Verlag, New York and Berlin (1984).

[5] Matsuoka O., Yoshimine M. and Clementi E., J. Chem. Phys. **64**, 1351 (1976).

[6] Corongiu G. and Clementi E., Gazz. Chim. Ital. **105**, 273, (1978).

[7] Kopka M.L., Fratini A.V., Drew H.R. and Dickerson R.E., J. Mol. Biol. **163**, 129 (1982).

[8] Corongiu G. and Clementi E., Biopolymers **20**, 551 (1981).

[9] Ranghino G., Wipff G., Wurtz J.M., Moras D., Thierry J.C., Westhof E., in preparation.

Round Table Discussion on

THE ORGANIZATION OF A MOLECULAR MODELING GROUP

IN A CHEMICAL INDUSTRY

Chairman: Professor Giuseppe Allegra, Polytechnic of Milano

C. TOSI

Let me say a few introductory words. Yesterday morning, in my opening speech to this workshop, I gave you the reasons which induced me to organize it. Now I would like to explain in more detail why I thought of a Round Table like this as the concluding act of the workshop.

From the presentation, by my collaborators of the molecular modelling team, of our recent work you have obtained an idea of what our current interests are. At this point, I believe, it is legitimate that not only you, but we too, ask a few questions: is the level of our work comparable to the most advanced standards of international research in the field of computational chemistry? are we moving in the correct and most up-to-date directions? are all the important subfields of this discipline suitably covered?

If the answer to at least one of these questions is negative something has to be changed. Any ideas put forward by our guests in this round table are welcome, and should help us make the correct decisions about the best strategies for improving our performance.

Anyway, it must be remembered that our kind of activity has to cope with a strong constraint, namely that the number of persons dealing in computational chemistry as their full-time job in a single centre is limited, and they, however well prepared may be, clearly cannot master the whole knowledge, even in a restricted area.

M. MARSILI

I have probably had an experience similar to yours, although I think I come from a more semantic approach to computer chemistry: this because I worked eight years with synthetic design programmes, in which the question concerning conformational energy is so far not so important, because such programmes still operate mainly on topological structures. Molecular topological structures are easily connected to chemical symbols. If you see a hexagon with a ring inside, like an Egyptian hieroglyph, then you see a symbol for benzene. Nobody is going to count the electrons in the benzene ring, but for the chemist's eye that is benzene, and you know by heart the reactions that benzene can make.

So, the first years of computer chemistry, the 70's, were completely dedicated to training a computer to deal with chemical symbols. On the other hand energetics of molecules, like conformational energy, were studies preceding this phase: Allinger and other researchers prove it.

Let's leave quantum mechanics out of the discussion, we do not deal with quantum mechanics; but empirical force-field calculations yielding internal molecular energies were made long before symbols and semantics were blooming in computer chemistry, especially in the initial steps of drug design.

So, we had two different things going: the first was an artificial-intelligence oriented approach, which has now produced ripe fruits that can be collected, and the second was the approach carried out by people like you, and others, who did excellent work in creating models and algorithms for describing a molecule in its three-dimensional aspects, in its energetical hypersurface for example. Now, in my view, the time has come where both branches of development can be joined. Symbols alone cannot fully provide the results that we expect, energetics alone, without strategies relying on chemical symbols, is not so complete as it could be. By bringing both efforts together we can provide for a global molecular modeling system.

If one wants to make predictions about the reactivity of a molecule, say a metabolic prediction, he needs a semantic program like a forward-search reaction modelling program. Its performance can be enhanced by knowing what energy is involved in the reaction. In knowing the conformational energy you can do stereochemistry, and give answers to questions like "is this kind of elimination possible or not?". I am not a specialist of molecular energetics as you are, so the question, whether your work is comparable to the other results in the world or not, cannot be correctly answered by me. However, I have the feeling that it certainly is.

It is advisable for your group, especially for your group, which leads scientific research in an industrially large organization like Montedison, to add to the route that you opened yourself in the jungle of exploratory computer chemistry, the 70's and the early 80's, to add something coming from other research philosophies, realized more with symbols and semantics.

You will, for example, find in the world excellent programs

for the prediction of molecular structures from the interpretation of spectral data. They have no quantum mechanics, nothing, just symbols, and symbol-managing structure generators; or synthesis design programs and systems for reaction retrieval.

One needs a large number of descriptors to model a molecular structure, and, in addition to energetic considerations, one could add dipole moments, reactivity of bonds, partial charges, polarizabilities and so on. This is what, in a much more reduced frame than yours, I am attempting to attain. I mean, you have a large operational staff around you, much bigger than the one I can activate at Tecnofarmaci; there are only three people, plus a hardware specialist, a computer scientist. We try to work our way humbly and slowly, sometimes crowning it with success, striving to bring forth advances in the computer treatment of chemical symbols and, at the same time, the physico-chemical description of the molecular world. We feel that this is what the Tecnofarmaci consociates, our industrial partners, want from us.

We feel that going one way only, whatever way you choose, is not enough to tackle real problems, the real problems we are confronting every week. My effort is to design a complex symbol-dependent system of interrelated programs, the SUPERNOVA global molecular modelling system. But I shall include many other programs, of typical calculational nature.

I would suggest that something should happen in your group, once that you have the need to include, say, synthesis design problems; then you will interact with people of this field. Or, if you must start solving difficult problems in drug design, where conformational energy is not a sufficient parameter to give a complete solution. Or, if you do chemometrical investigations, you will need programs that provide you with a larger variety of molecular descriptors, than the ones that you have presented here.

G. ALLEGRA

If I might make a couple of comments, first of all I think that there may be an enormous variety of quantities that may be calculated by the molecular modelling techniques we have heard about these days. But what I am considering now in particular are two possibilities. One is that, just listening to Dr. Fusco's presentation concerning the visualization of molecules in slightly different conformations around the energy minima, I see the possibility that those results could be checked versus the results of X-ray crystallography, which nowadays are so accurate in terms of thermal vibrations. It is rather intuitive that these spatial locations of the molecule may be considered as a kind of mapping of its molecular vibrations, which in turn may be detected by X-rays. It would be proper to add the bending contribution, which cannot be ignored in general. Anyway, as far as I can say, this should be a very fruitful direction to move in, because you really have a mine of possible experimental information, which is becoming increasingly accurate nowadays.

Another consideration I would like to propose is that these techniques of molecular modeling could be very fruitful in testing some formal treatments of conformational statistics of long molecules. I am now referring to the presentations by Scordamaglia and Mrs. Barino yesterday concerning the mapping of the hypersurface of conformational energy of a macromolecule. Are you now thinking of the synthetic macromolecules in which you have a repeating unit which is pretty short? In this case, you know that there are nowadays well-developed techniques with correlation matrices, which predict the entropy, the energy and the dimensions of the macromolecule. Your method is in principle very rigorous because it is based on a sound statistical approach. It could provide a very interesting check on whether those matrix methods are really accurate in describing the macromolecular conformations. I think that you might for instance suggest that those matrices should be expanded to account not only a correlation between neighbouring rotational angles, but perhaps between one pair and the next pair, or something of that sort. In other words, you might check the Markoffianity of your system, i.e. the chain, in terms of the rotational states, which is a critical current assumption. And that is a very interesting sort of test, I believe, because not much has been done in trying to assess really the degree of accuracy of those matrix methods. And I believe that if you can do that, this would be a very important scientific step.

C. TOSI

I would like to add a comment about Allegra's remarks on the presentation by Fusco on methotrexate. We have obtained some more results about this molecule, which have not been presented here, because they were already presented by myself in Budapest at a congress of theoretical organic chemistry in August. One seems to be very interesting and to have far-reaching consequences, namely that there exist in literature three crystal structures of this molecule, one as a single crystal and the other two bound to a protein, the DHFR. Well, all of these are high-resolution investigations; but, if you consider the small molecule, high resolution means that you know the atomic positions to an accuracy of ca. 0.05 Å, while high resolution for a protein means 1.8 to 2 Å in the most favourable cases. Actually, we have observed that, with the same force field, the energy of the single-crystal structure is 3 kcal/mol above the global minimum we have found, while the energies of the two structures bound to the protein are ca. 50 kcal/mol above the global minimum. Now, it is possible that this be the consequence of the bad resolution, but I am far from being sure that, by modifying somewhere the internal coordinates, you might be able to cause the energy to decrease by so large an amount; so, it is possible that the presence of a number of hydrogen bonds with residues of the protein allows the molecule to have an internal strain much, much higher than it was thought in the past. And, for example, a very interesting question to be answered would be this: if we consider the pocket of the protein structure within which the inhibitor molecule is inserted, is it possible to minimize the energy of this molecule, starting from the crystal conformation, in such a way that it reaches the minimum-energy conformation consistent with the constraint of remaining within this pocket?

G. ZERBI

I do not know what your problems are, what kind of internal interactions you have, in which way you choose your problems, and how you depend on your local internal structure. I think you are mastering a beautiful way of calculating structures and predicting the properties of your molecules. It could be of great benefit to many people if you would consider what problems can be solved through your modeling in other fields such as NMR, mechanical properties, and so on. My feeling is that you could give great momentum to your kind of research, if you offer other people in your company help from theoretical calculations.

J.-M. ANDRE

I have just one comment to make in the same field as Prof. Zerbi. When you speak of two different fields, I mean computers in chemistry for example, it is always necessary to know the emphasis you put on both sides of the problem. And that has always been the case in physical chemistry or in chemical physics; for example, there is a strong difference between both fields: when you see the Journal of Physical Chemistry, that is not the Journal of Chemical Physics. That is the same problem, I believe, for computers in chemistry, and if you speak of computers, I believe that in this Institute you have the correct way of thinking. You have the semi-empirical programmes, the empirical programmes, the effect of temperature, Monte Carlo calculations, molecular dynamics, and so on. And on the other side you have the problems which are posed by the industry. So, the basic difficulty is (i) to select the problems to be solved, and (ii) to select the correct way or the adequate methodology for solving them.

R. MOLL

I want to give a short personal impression from five weeks of work here in the group of Prof. Tosi, and first I want to give a short comment to the words of Prof. Marsili. I agree with him completely, because the situation of the group of Prof. Tosi and of Donegani is the same as the situation in our company, only the points are quite different. In Donegani a group of molecular modelling exists and no synthesis design, and in our company we have synthesis design and no molecular modelling. I think, both have to establish the missing parts. As to the staff problems here in Donegani I cannot speak. But Dr. Guaragna and I have made some proposals and recommendations for the introduction of computer-aided synthesis planning in this Institute, and we will discuss these in the next few weeks.

Furthermore I have the impression that in the group of Prof. Tosi there are five specialists, and five excellent specialists in their field, but all of them work on conformation analysis. In my opinion this is, maybe, too strong a reductionism to one complex problem. It is very difficult for me to advise or to say, what has to be done in this situation, because there are, among other things, limitations in staff, funds and so on. We have similar problems in our company. We are also in the position of having some specialists, e.g. in pesticide research, and we have to cover a rather broad field, but we can do only a small part of the work we want to do. Maybe we will have the opportunity to discuss our problems in the next few weeks and, it is hoped, we will come to some results of benefit for both our groups and for both our companies.

D.P. DOLATA

Let me be slightly controversial here, and see what we can kick-up. Montedison is a production company. Its goal is to produce something which they can sell and get back money for. Now, that means you can either contribute to produce chemicals that will produce a desired result in the world, or you can produce software which will be sellable. If you are not trying to produce software, maybe you should go for the slightly more conservative approach, you should not try and be on the very cutting edge of software research because you can make mistakes: I have made a lot of mistakes, I have wasted many years in doing things that, when I look back at them, were stupid. Now, that's fine when you are an academic, because as an academic you are trying to sell software, and new ideas, and exploratory things. But I am taking a big risk, and Donegani and Montedison are not in the business of taking really huge risks, they are in the business of doing things and doing them well. Now, another thing:

as was mentioned, you have these people all seeming to work more or less in the same area. If you are going to sell chemicals or products that do something, you need to be able to have computer programmes which help you identify the action you want, help identify a chemical which will produce that action, help identify the context which will produce that action, help identify methods that will synthesize that chemical, help identify what to do with the side products and things like that. It seems to me there are a lot of programmes out there which do all of that, and you should be mining the academic environment to try and bring those programmes in here. You might be a year or two behind the really most advanced work, but that might be the way to get a really solid beginning to an end computational result, which will provide the product that this company needs.

J.-M. ANDRE

I would like to comment on Dolata's answer. I believe that at the European Community (EEC) there are at the moment some programmes for connecting universities and industries, and industry institutes, and that should be investigated a bit further. They are excellently designed to push the cooperation between industry and universities and to favour a close collaboration between scientists of different specialities.

M. GIL

Just a piece of information from Prof. Tosi. How is your group organized? How many people are working? What kind of facilities do you have in your molecular modeling group? if it is possible to know.

C. TOSI

There are six of us altogether, including myself. Three are physicists and three are chemists. Often the choices are the result of historical circumstances, more than of decisions taken in advance. And we have had here, you know, a department of chemical computation directed by E. Clementi. This was a very great breakthrough at that time. But together with a lot of success and of advantageous consequences, it has also had some negative consequences, namely, that maybe the gap between the mentality of those who are doing this work and of those who should benefit by its results is increasing instead of decreasing. It is very difficult to dialogue with other people, because there is a general feeling that this is a magic instrument that can give an answer to everything. For many people it is difficult to realize that it is not the fact that one masters a certain method that enables him to obtain good results, but the fact that he is able to design a good model for a certain problem. In other words, I believe that the most important problem is not to apply a method, but to decide in advance how to model, how to transform the request of some people in the Institute into a problem which can be performed in a logical way, and which is able to give a real answer to his needs.

G. ZERBI

I think we have to be realistic. Each one of us is interacting every day with friends in organic chemistry or engineers; everybody comes up and says "I would like you to do that". Most of the times the questions they raise are totally pointless for us. I am speaking from my own experience. Generally, they do not know what you can do and what kind of problems you can solve. If you try to tell them what your tools are, and where the usefulness and limitations of your tools are, people approach you in a different way. It is my own personal experience. Organic chemists know nothing about computational spectroscopy. They ask crazy questions, which are unsolvable, and if you do not please them, they say you are useless. The engineers ask you to study fracture, but we will never be able to study fracture well with our toys. Everyone of us is in touch with an environment which does not know what we are doing; that is normal in everyday life in every laboratory throughout the whole world. The only way to get a breakthrough in your environment is to tell your friends what your tools are, what are they used for, and try once to show that their problems can be solved thanks to your help. And this is the way you can eventually gain more momentum, more money, more people. I do not know how things are here. This is the way everyone of us is facing, in our own survival for research.

D.P. DOLATA

Yes, that sales can take on several aspects. The SECS programme (Simulation and Evaluation of Chemical Synthesis), written by Prof. Wipke, is used by several companies in Germany and Switzerland. One of the things they found out was that this programme was labelled as artificial intelligence, and is supposed to be able to create synthesis for molecules. The only way that they could get laboratory chemists to regularly use the programme was to lock it away in a room, allow the chemist to go in, play with the programme, come out, several days later come up with the synthesis and not have to attribute who came up with which ideas. Because the chemists find the idea of saying "Well, this synthesis was devised by the programme and I only pushed the buttons" very threatening to their intellectual abilities, to their job future and things like that. So, one of the sales aspects is that we have to produce tools for the chemist rather than replacements for the chemist, or chemists will not want them.

G.B. BACHELET

As an outsider of both the chemical community and of the private-industry research in Italy, and a former employee of AT&T Bell Laboratories and Max-Planck Institute, I would like to make a remark which is rather in contrast with the previous one.

My limited experience suggests that not only in Italy, but more generally in Europe, the goal of efficiently combining research and development (R&D) is seldom achieved, the main problem being the lack of critical mass; the diffuse skepticism towards long-range R&D policies is probably another important factor. Typically in the United States one encounters gigantic corporations, which can afford very large R&D laboratories with maybe a few thousand employees. In such a context fifty or a hundred scientists are hired to do pure research; their role is to provide a stimulating environment where good ideas are born and discussed, with no immediate link to the production lines. The investment is considered to be a highly rewarding one, because, beyond a given "critical mass", the interaction of a cluster of pure scientists with the rest of the laboratory is known to be a key factor for the promotion of a creative environment and ultimately for the supremacy in the "gold rush" of technological innovation. Moreover, since this policy of large industrial corporations turns out to be very successful in terms of patents and technological breakthroughs, even medium-small companies, where the "critical mass" cannot be reached, do often hire a few pure scientists in their labs in the hope of keeping up with progress and having a chance in the technological race. The situation on our continent appears radically different. Even corporations which are experiencing a commercial boom, like Olivetti in Italy, have neither the critical size to afford large research labs nor the American philosophy about the potential industrial impact of basic research. Their best minds are often forced to concentrate almost exclusively on development, with little space and no encouragement for basic research. As a result, basic research develops almost entirely within Academic or Government organizations.

My impression is therefore that Prof. Tosi's group here represents one of the few miracles in the area of private research in Europe, certainly in Italy. This is probably also due, as Prof. Andre was saying, to vital links between university and industry, which were promoted by CNR and other European projects. It appears that the potential of a few top-quality scientists is here allowed to be invested in pure research, even if this effort may not immediately produce, let us say, an increased volume of sales for Montedison. My wish is of course that this remarkable experience be continued in the next years. I would like to ask Prof. Tosi two questions: how could he find this space in a private industry before now? Were (and are) there any difficulties in keeping the balance between company business and research?

C. TOSI

As I already mentioned, our Molecular Modeling Group is the heir of the Department of Chemical Computation established at Istituto Donegani by Enrico Clementi in 1974. The birth of that department, in turn, was the fruit of the far-sightedness of the former Director of the R&D Division of Montedison, Professor Umberto Colombo, who had perceived the importance of the computational approach as an aid to solving problems in modern chemical research. Unfortunately the success that enterprise met was not as large as it would have deserved, and the consensus to it was, by many people, more formal than substantial.

In the late Seventies Clementi was one of the first to foresee the spectacular development that computer architectures would have in the coming years, and the Istituto Donegani, where application of computer techniques to problems of practical concern is more important than improvement of software, not to say hardware, was not the best suited place for him to be a leading actor in such a development. Essentially for this reason he went back to the United States, and the computational chemistry activity continued here, on a reduced scale, under my leadership. Obviously the credit our managers give to our activity, and the resources they accordingly grant to us, are not an indefinitely vested right, but something we must earn day by day, by convincing our management with deeds, not words, that they are not wasting money. Our persuasion work is the easier, the larger the contributions we give to helping our colleagues in solving their problems and in looking at them in a more rational way.

It is only the closing of the cycle {initial credit - assignment of resources - achievement of useful results - renewal of credit and resources} that warrants our survival.

D.P. DOLATA

Let me just respond in favour of "prostitute" research, i.e. research for money. I feel that in a way, building axiomatic theories is a very pure field of research. But until I actually brought it down and got it into the laboratory at Oxford, and got it to a point where it was being useful for an actual chemical reaction, it was just building castles in the air. And, when I actually got my research results down to the nitty gritty of the laboratory, I got something which was useful to me, a product which is a research paper. But that was to me very valuable, and I think that somehow people, especially mathematicians and then physicists and then chemists (and we cannot forget biologists), have this idea of "pure", as being "pure". Maybe we should change terms to "research without direct applications" and "research which is useful". And then if you change the titles from "pure" to "research without applications" and "applied" to "research which is useful", suddenly it seems that it is much more beautiful to do applied research than pure.

G.B. BACHELET

The point is that usually you do not exactly know from the start which research is going to be absolutely useful and which absolutely useless. Otherwise it would be trivial to make R&D policies in the industry. The experience of some major corporations (like AT&T or IBM for example) is in fact that many of the important findings, which eventually turned out to be of key commercial impact, were not born in the development areas, but in the "pure" research areas of their laboratories. That is why they continue to pour billions of dollars into "pure" and apparently useless research.

Two examples. The recent discovery of high-temperature superconductors occurred at the IBM Research Laboratory of Rüschlikon (Zurich, Switzerland); there many independent solid-state scientists interact in a stimulating atmosphere where theoretical and experimental innovation is clearly predominant over industrial pressure for new devices. The transistor itself, as well as other major achievements in microelectronics, were conceived in the research areas of AT&T Bell labs, Murray Hill (New Jersey, USA), where care has always been taken to keep a significant group of "pure" experimental and theoretical physicists and chemists free from the pressure of short-range applications, so that they can "think big".

The emphasis on industrial achievements is far from being absent in the two labs just mentioned; one can still breathe it all over the place. And in fact, to complete my picture, I should add that some pressure towards industrial applications turns out to be a highly stimulating factor on pure science; it is really the interplay between a large group of applied scientists, who are focused on well-defined industrial goals, and a small group of pure scientists, who are allowed to do almost whatever they want, which seems to be the rule - seldom followed in Europe - for a successful R&D laboratory.

D.P. DOLATA

The transistor was developed in Bell labs in a totally pure environment and within several years after its development, the entire world was free to utilize it and has been doing so, thank you very much. High-temperature superconductors are being explored everywhere. Those places which have enough money to do this basic research seem to be pretty interested in giving that information out to the whole world. And so, when you have a limited resource, which Montedison and Donegani have, you could hope to be able to make this once in a lifetime, Nobel discovery, but you probably should recognize that the percentage chances on that are not going to pay. Unless you really consider public relations, that is valuable. And if I were directing a company like this, I would probably (and here is where Fusco hits me) put my people more into the applied, one or two years behind the really state-of-the-art work, just to make sure that there is enough time for the wrinkles to work out. I am sitting up here and I am talking about "expert systems". Two years ago everybody was saying that they are the neatest things since sliced bread. Now people are beginning to realize that they do not do everything. There is about a two-year time lag between the sales pitch and when people started admitting that it is not so wonderful. And it is a good thing to be able to find out that it fails before you invest one or two man years, person years.

G. ZERBI

Referring to your previous remarks, I wish to express some optimistic wishes to the group here. Since I am older than you, I am somewhat skeptical about large groups working together. Even if you look outside the small Italy, large groups are always made up of smaller groups which push. The best contributions come from small groups with bright ideas working hard day and night, with post-docs not well paid to do a lot of work. The more people you have, the more human problems you have, and it becomes harder to justify your budget to the bosses of the company at the end of the year. I think that six to ten people really working together make an ideal team to get something done. And this is what I wish for the group of Prof. Tosi. The brightest ideas do not come from large coordinated groups which do not exist, even in the big United States. Even during the final rush for the high-TC case, a few people worked day and night, and then came up with an high temperature superconductor in a very artisan way; no plans, no large authorizations, no signature by any boss, just make it and then we'll see. I do believe that since Donegani still cares for basic science, your group has a chance to become useful to the other groups within your Company. The more people feel useful the happier they become.